RAND

Proceedings of the RAND Project AIR FORCE Workshop on Transatmospheric Vehicles

Daniel Gonzales, Mel Eisman,
Calvin Shipbaugh, Timothy Bonds,
Anh Tuan Le

Prepared for the
United States Air Force

Project AIR FORCE

Preface

Transatmospheric vehicles (TAVs) are envisioned as a new type of reusable launch vehicle that could insert payloads into low earth orbit or deliver them to distant targets within minutes. Such a vehicle may be able to carry out several types of military, civil, and commercial missions. In past decades, a number of military TAV concepts have been proposed, but a complete operational vehicle has never been built.

The promise of TAVs is that because of their reusability they could launch payloads at much lower cost than existing rockets. In addition, if they were operated more like aircraft and less like rockets, they could enable responsive and flexible space operations, features that would be useful for a number of military missions.

A workshop was held at RAND on April 18 and 19, 1995, to examine TAV mission, technical feasibility, and design issues. Experts from the Department of Defense, the National Aeronautics and Space Administration (NASA), and industry participated. This report summarizes the proceedings of that workshop and subsequent research into questions raised at the workshop. We examine potential missions and their implications for vehicle design, cost, and size. A variety of contractor TAV design concepts were presented and discussed, including designs currently being developed for the NASA X-33 and X-34 programs.

This work was done for the Future Role of the Air Force in Space project, one of several studies being carried out in the Force Modernization and Employment Program of Project AIR FORCE.

Project AIR FORCE

Project AIR FORCE, a division of RAND, is the Air Force federally funded research and development center (FFRDC) for studies and analyses. It provides the Air Force with independent analyses of policy alternatives affecting the development, employment, combat readiness, and support of current and future aerospace forces. Research is performed in three programs: Strategy and Doctrine, Force Modernization and Employment, and Resource Management and System Acquisition.

Contents

Figures

Tables

Summary

Launch vehicles or rocket boosters are used to deliver satellites to orbit or weapons to distant targets. However, most existing rockets, based on designs over a quarter-century old, are expended after use, making them less cost-effective for some missions and less competitive in the world launch market. A transatmospheric vehicle (TAV) or reusable launch vehicle (RLV) would be capable of returning to earth to be reused after refurbishment and refueling.[1] TAVs may have aerodynamic and operability characteristics similar to conventional aircraft, but would be capable of delivering payloads to low earth orbit (LEO). The promise of TAVs is that their reusability would potentially allow them to launch payloads into orbit at much lower cost than expendable launch vehicles.

Purpose of the Workshop

The Air Force asked RAND to examine the utility, feasibility, and cost of procuring a TAV capable of carrying out military missions. As part of this study, a workshop was held at RAND on April 18 and 19, 1995, to examine TAV utility, technical feasibility, and design issues. Experts from the Department of Defense, the National Aeronautics and Space Administration (NASA), and industry participated. A variety of commercial RLV and military TAV design concepts were presented and discussed, including designs submitted for the NASA X-33 and X-34 program competitions.

The focus of the workshop was to determine whether any technology "show stoppers" exist today that could seriously impede development of a TAV. Although several technology risk areas were identified, recent technological advances suggest there are reasons to be optimistic that a TAV with significant military utility could be developed.

Discussions held at the workshop raised a number of issues that we believed warranted further investigation. These subsequent ancillary analyses, included in this report for completeness, delayed publication of the proceedings.

---

[1]We will refer to civil or commercial reusable launch systems as RLVs, and to reusable launch systems designed for military missions as TAVs. Recently TAVs have also been termed spaceplanes.

The results summarized here are subject to the following caveats. Proprietary presentations given at the workshop have been made nonproprietary to enable the widest possible distribution. Thus, the reader may find some important but sensitive pieces of information missing. In addition, the information cut-off date for this report is April 1995, with the except of data on X-33 vehicle concepts. Further analyses on some of the TAV concepts discussed at the workshop and new TAV or RLV concepts are not included in these proceedings.

Lessons Learned from Previous Programs

In the past few decades, a number of military TAV concepts have been proposed, and needed critical technology development work has been conducted, but no complete functional vehicle has ever been built. The first National Aerospace Plane (NASP) program, which took place from the late 1950s to the early 1960s, demonstrated a number of important technologies such as real-time air liquefaction and operations such as hypersonic refueling that could be important in Two Stage To Orbit vehicles. The second NASP program in the 1980s was equally ambitious. The later NASP design called for a Single Stage To Orbit (SSTO) fully reusable system based on a complex combined cycle propulsion concept that had several air-breathing engine components. Because of the high technological risk associated with this propulsion concept and other vehicle design aspects, NASP never proceeded beyond the technology development phase. The program was canceled after $1.7B was spent and it became clear that an operational prototype would have cost $10B or more. In light of the difficulties encountered with NASP, only TAVs based on existing or demonstrated jet and rocket engines were considered at the workshop.

Potential RLV and TAV Mission Needs

Potential RLV and TAV mission needs fall into civil, commercial, and military categories.

Potential Civil Needs

NASA's space launch needs can be divided in two categories: launch of satellites for environmental monitoring and planetary exploration, and launch of astronauts and payloads for the U.S. manned space flight program. NASA's most important need from a financial point of view is to replace the space shuttle with a follow-on man-rated vehicle. A shuttle follow-on could be smaller than the current system, but it may have to have a heavy lift capability to support the space station logistically or future expansion or refurbishment of the space station. In either case, NASA will increasingly be motivated to develop a shuttle follow-on RLV with lower operating costs, as the shuttle consumes a large portion of NASA's current budget. Because an RLV may be needed in the 2005–2010 timeframe and because it may take a decade or more to develop, NASA has already embarked on an R&D path to do this by initiating the X-33 program.

Commercial Needs

Over a decade ago, U.S. companies dominated the international space launch market. Today, Arianespace with its line of Ariane launchers controls over 40 percent of the market. China and Russia have also aggressively marketed their launch vehicles. Restrictions on the price and number of Russian and Chinese launches have so far prevented serious market disruption or financial injury to U.S. launch vehicle providers. However, these cartel-like arrangements may not stay in place indefinitely.

Today, U.S. launch vehicles either cost more or have smaller payload capability than do foreign competitors. The challenge faced by U.S. industry is to meet the low launch prices currently offered by Russian and Chinese launch providers. An RLV developed and owned by U.S. industry may be the best way to meet this competitive challenge.

Most commercial satellites today are medium-sized, and most commercial satellites are launched by medium launch vehicles (MLVs). Consequently, an RLV designed for the commercial market would most likely be an MLV.

Potential Military Needs

Potential military TAV needs can be grouped into three mission areas: space launch support, space control, and space force application.

Space Launch Support. Many current and planned DoD payloads require an MLV launch. These needs could be satisfied by expendable boosters or possibly by a commercial RLV.

In the future, the DoD may rely more on commercial and civil satellites. However, it may not be possible to optimize the regional coverage supplied by commercial or DoD satellites that are stored on orbit far in advance of hostilities. However, optimal regional coverage can be obtained if "gap filler" satellites are deployed rapidly at the onset of crisis or conflict. A highly responsive small launch vehicle (SLV) or a TAV could quickly deploy such "gap filler" satellites.

Space Control. The United States may require new space control capabilities in the near future to counter increasingly capable foreign and commercial satellites. A highly responsive TAV could rapidly deploy space control payloads during crisis or war. Recent RAND research indicates that a variety of useful space control payloads would be small enough to be launched by an SLV or TAV.[2]

Space Force Application. The development and employment of space weapons or the delivery of weapons through space would first require a national decision. However, there are reasons to consider acquiring such capabilities: to counter or deter ballistic missile attack or the invasion of allied countries by enemy ground forces, and to attack heavily defended high-value or very threatening targets in hardened facilities or deeply buried bunkers.

If appropriate weapons were available, a highly responsive TAV could attack ballistic missile launchers within minutes after TAV launch. Such a capability could have significant deterrent value and could provide a global counterforce capability. Similarly, if a TAV could quickly deliver such weapons against terrestrial targets such as armored vehicles, it could serve as a deterrent to regional aggressors or slow invading armies, perhaps within minutes after the border was crossed and before allied cities or industrial facilities were captured.

Another advantage of TAV-delivered weapons is the high kinetic energy such weapons can impart to a hardened or deeply buried target, enabling increased target penetration and weapon lethality.[3] A TAV capable of delivering such weapons against hardened and heavily defended targets would add an important new capability to the U.S. arsenal.

Launch Vehicle Responsiveness

Responsiveness is one of the most important attributes of a military TAV. We define launch vehicle responsiveness as the time needed to prepare a new vehicle or one that has just returned from space for launch. Nominal responsiveness estimates for current U.S. launch vehicles versus their payload delivery capabilities to LEO are plotted in Figure S.1.

From the figure it is apparent that current medium and heavy lift launch vehicles are not responsive and that vehicle responsiveness improves dramatically with decreasing payload delivery capability. Consequently, existing U.S. launch

---

[2]Further discussion of this subject is beyond the scope of this report.
[3]Sandia Laboratories, unpublished research made available to RAND.

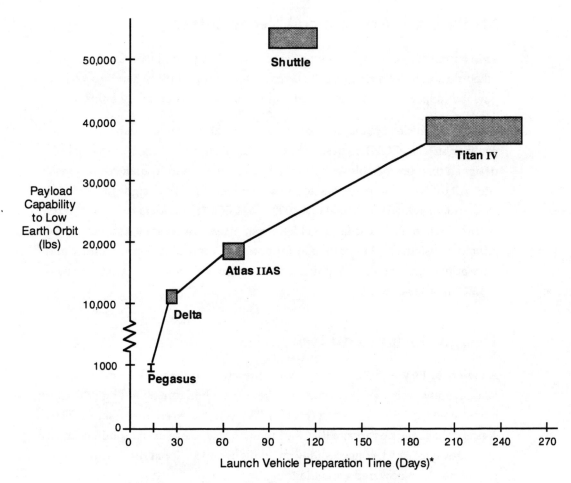

*Excludes time delays caused by launch vehicle failures

Figure S.1—Responsiveness and Payload Capabilities of U.S. Launch Vehicles

vehicles, with the possible exception of Pegasus, could not support the timelines required to effectively carry out the military missions described above.

However, if a TAV could be launched within minutes or hours of a launch order, it may be possible to effectively carry out these missions. This degree of responsiveness may be possible only if TAVs could be operated like aircraft and be put on alert like bombers. Aircraft-like levels of responsiveness imply aircraft-. like levels of supportability and reliability. Aircraft-like supportability in turn implies a much higher level of reliability than that of current launch vehicles. The data in Figure S.1 suggest that an operationally useful military TAV should be built to handle as small a payload as possible in order to maximize TAV responsiveness.

Military and Commercial Needs Differ

The combination of a rapid launch-on-alert capability, unpredictable launch schedule, fast turnaround time, and rapid reconfigurability to handle a variety of payloads appear to result in a set of requirements that is uniquely military.

A military vehicle capable of being launched on alert from a number of continental U.S. (CONUS) bases could be very different from a commercial RLV designed for a predictable launch schedule and operation from only one launch site. All the X-33 competitors sized their commercial RLV designs to handle substantial payloads of 20,000 to 45,000 lb to LEO. If history is any guide, it will be difficult to make these large vehicles responsive. In contrast, for many military missions, and in particular for potentially important space control and force application missions, a payload delivery capability of only 1,000 to 5,000 lb to LEO may be adequate.

Design Options and Issues

A variety of TAV and RLV designs were presented at the workshop. Vertical take-off and landing (VTVL), vertical take-off and horizontal landing (VTHL), or horizontal take-off and landing (HTHL) TAV concepts were discussed. HTHL concepts have the advantage that they may be able to use existing infrastructure and runways. VTVL concepts may entail higher operational risk because of the requirement for powered rocket landing.

Vehicles also came in a number of staging concepts, including SSTO, Two Stage to Orbit (TSTO) air-launched, and TSTO aerial-refueled concepts.

Table S.1 lists most of the RLV and TAV design concepts presented at the RAND TAV workshop. Several of these designs are based on detailed technology and design studies, such as the X-33 entrants, while others reflect promising but newer and less thoroughly explored concepts. The level of maturity of these concepts is indicated by the legend in the table.

Several TAV concepts presented at the workshop have been under active investigation in DoD, including the Black Horse concept. Subsequent to the workshop, RAND performed an independent analysis of Black Horse's payload capability and found it could not reach orbit. RAND performed a similar analysis of the Northrop Grumman (NG) TAV and was able to verify the contractor's claimed payload delivery capability.

The NG TAV concept appears promising and could be well suited for several military missions. It could potentially deliver a 1-6 klb payload to various LEO

Table S.1

Recent RLV and TAV Concept Proposals

Vehicle	Contractor/Lab	Staging	Payload	Propulsion	Comments	Status
X-33	Lockheed Martin	SSTO	Heavy	LOX-LH2	Lifting body, VTHL, aerospike engine	■
X-33	Rockwell	SSTO	Heavy	LOX-LH2	VTHL	▨
X-33	McDonnell	SSTO	Heavy	LOX-LH2	VTVL	▨
X-34	OSC	Air-drop	Small	LOX-storable	HTHL, L-1011	■
REFLY	Rockwell	Air-drop Pegasus	Very small	Noncryogenic	L-1011, B-52, reusable upper stage	□
NG TAV	Northrop Grumman	Air-launched	Small	LOX-LH2	Boeing 747	☑
Black Horse	Phillips Lab	Aerial-refueled	Small	H2O2-kerosene	KC-135Q tanker	☒
Neptune	Phillips Lab	Air-drop	Small	LOX-LH2	B-1B	
TAV	AMC HQ (Snead)	Air-launched	Medium	LOX-LH2	Boeing 777	

■ Under development (NASA) ▨ Design proposed □ Concept proposed

☑ Concept performance verified ☒ Concept performance problem identified

orbits and would be launched from on top of a Boeing 747. The orbital vehicle resembles a scaled-down Space Shuttle and would have its hypersonic characteristics and a significant cross-range capability.

Relatively little design and development work has been done on TSTO TAV concepts when compared to the work done on SSTO concepts. A broader set of air-launched and aerial-refueled TAV concepts deserves study, including study of high-density propellants and air-launch separation dynamics at supersonic and subsonic speeds.

The X-33 competition, much like the NASP program, has focused attention on SSTO vehicles. RAND believes TSTO TAV concepts deserve equal attention if delivering small- to medium-sized payloads to LEO is viewed as a primary mission need. Air-launched TSTO TAV concepts appear particularly promising from a cost standpoint because the first stage aircraft could be based on a commercial civil air transport. In addition, they may provide an evolutionary development path to full reusability and aircraft-like levels of responsiveness for orbital vehicles. In contrast, SSTO systems may be more challenging technically, much more costly to build, and would be so large they could not meet military responsiveness needs.

Technology Challenges

Developing a highly responsive and cost-effective TAV regardless of the design approach chosen will be challenging, but is certainly possible given the advances made in key technology areas in the past few decades. Much remains to be done, especially in propulsion.

Advanced Materials and Structures

Minimizing vehicle empty weight is important for any vehicle concept and critical for SSTO concepts. This will require the integration of lightweight composite materials into the vehicle airframe and subsystems. Modern composite materials have higher strength and stiffness than standard metals, which can significantly reduce overall vehicle structural weight. For example, per-unit-weight graphite epoxy is five times stronger than aluminum alloy, the material the space shuttle airframe is composed of. According to some analyses, advanced composite materials and lightweight metal alloys may permit launch vehicle structure weight to be reduced by up to 35 percent.

Propulsion

A TAV's rocket engines will have to be efficient and provide the thrust levels needed to reach orbit. Rapid turnaround and low-cost operations also require that engines be durable, damage-tolerant, easily inspectable, and capable of rapid and safe shutdown.

From performance and reliability standpoints, the best bi-propellant engines use cyrogenic liquid hydrogen (LH2) and liquid oxygen (LOX). However, cryogenic fuels introduce handling complications that may reduce TAV responsiveness. Hydrogen leaks are especially difficult to contain, pose an explosion hazard, and introduce additional operability concerns. Further research is needed to find ways to efficiently handle LH2 in a military operations environment.

High-density propellants are not as efficient as cryogenic propellants, but they may provide operability advantages that could be especially useful for military TAVs. In addition, vehicles based on high-density propulsion could be significantly smaller in size, which could provide special advantages for TSTO air-launched concepts. Research is needed on high-density propellants to determine whether these possible benefits can be realized.

The highest-thrust LH2/LOX engine available today is the Space Shuttle Main Engine (SSME). The current SSME does not have the performance levels needed for a full-scale SSTO RLV, and even planned improvements to this engine may not be adequate for this purpose, calling into question the feasibility of building an SSTO RLV with conventional rocket engines.[4]

The current SSME must be pulled from the shuttle after every flight to replace the turbopumps, although planned engine upgrades may reduce engine replacement time to once every ten flights. The Russian D-57 LOX/LH2 engine provides slightly lower performance than the SSME and is smaller, but has adequate thrust levels for the TSTO NG TAV concept and would have to be pulled from the vehicle only once every ten flights. This level of durability should be adequate for a military TAV. However, if even more durable engines can be built, TAV responsiveness could be increased and launch operations costs could be reduced further.

LOX/LH2 aerospike rocket engines, like those planned for the Lockheed-Martin X-33, may have significant performance advantages over conventional rocket

---

[4]The current Block II+ SSME has a thrust-to-weight ratio of 58, which NASA is planning to improve to near 70 in the Block III SSME. The National Research Council has reported an SSTO RLV would require a thrust-to-weight value somewhere between 75 and 80.

engines. They may be able to deliver nearly the same level of performance as the SSME, but have the higher thrust-to-weight levels needed for SSTO applications. This is an area where the Air Force and DoD could benefit substantially from work being done for NASA.

Finally, tri-propellants may offer performance advantages and higher densities, potentially leading to smaller, less-expensive vehicle designs. However, past investigations have been discouraging. Nevertheless, because of their potentially high payoff, tri-propellants warrant continued research.

Thermal Protection Systems (TPS)

Rapid turnaround between missions, cost-effective operation, and high payload mass fraction characteristics will also require development of a lightweight, robust, and durable TPS. Our review of TPS materials reveals that it should be feasible to design a durable TPS from advanced metallic alloys, provided the reentry path and the vehicle's aerodynamic design result in reentry temperatures that are less severe than those found on the space shuttle. Peak temperature locations would probably still require reinforced carbon-carbon to withstand reentry thermal loads, but most other locations on the vehicle should be protectable by combinations of metallic panels. Although metallic panels would have higher density than ceramic tiles, a metallic TPS may be lighter and simpler by eliminating the need for the complex adhesive system used on the space shuttle. The panels may also serve as aerodynamic load-bearing structures, eliminating the necessity for an underlying airframe.

Furthermore, by optimizing the vehicle's aerodynamic design, it may be possible to reduce the thermal loads on the vehicle, thereby decreasing the degree of thermal protection required. These improvements could result in a TPS that is more reliable and less expensive to maintain than that of the current space shuttle ceramic tile system that requires 17,000 man-hours for refurbishment after every flight.

Vehicle Integration

A significant engineering challenge is the effective integration of lightweight high-strength composites and metal alloys into vehicle structures and the integration of durable metallic TPS and efficient and durable rocket engines into the vehicle. A TAV that has these technologies and subsystems effectively integrated would likely have desirable payload delivery characteristics and could responsively carry out a range of military missions.

Multimission Capability and Cost

Cost is of course an important design consideration. A military TAV capable of supporting only one mission area, such as launching small satellites into LEO, may not be cost-effective, except possibly over the long term (in terms of total life-cycle costs). For a small military TAV, such as the NG TSTO TAV concept, a budget of about $760M would be required to build one subscale prototype X-vehicle and one full-scale operational prototype.[5]

A military TAV should therefore have a multimission capability to justify an expenditure of this size for TAV development in today's austere budget environment. Before any development decisions can be drawn from the above discussion, a thorough analysis of TAV mission cost effectiveness should be performed and the results compared with the capabilities of other platforms. Such a mission analysis is beyond the scope of this report.

Conclusions

TAVs could potentially launch payloads into space or toward distant targets at much lower cost than expendable launch vehicles. In addition, if they could be operated more like aircraft and less like expendable rockets, they offer the promise of carrying out space operations with much greater flexibility and responsiveness than is possible today.

Discussions at the workshop and subsequent investigations reveal that despite the efforts of past programs, significant technology challenges remain, especially in the areas of propulsion, thermal protection systems, and overall vehicle integration. Stringent mass fraction limits will have to be met for the vehicle to reach orbit with its intended payload. Overall vehicle design is very important. It is too early to know which sort of vehicle design has the best chance of meeting required mass fraction limits. More research is needed in propulsion, thermal protection systems, and overall vehicle design. The NASA X-33 program will provide important new data in all three areas, but the DoD needs to pursue research in all three areas as well.

A reusable launch vehicle could satisfy civil, commercial, and military space launch needs. However, our analysis reveals that civil and commercial launch needs differ in some important respects from emerging military needs. The highest priority for civil and commercial users is to reduce the cost of access to

---

[5]See Section 2 and *Life Cycle Cost Assessments for Military Transatmospheric Vehicles*, MR-893-AF, 1977, for further details regarding this cost analysis.

space. Military users are also concerned about reducing costs, but launch vehicle responsiveness and flexibility are critical for some military missions. These differing needs have an important bearing on vehicle design and imply that a military TAV may differ in important ways from an RLV designed exclusively for the commercial launch market.

Finally, reducing launch vehicle costs will at best address only half the problem of reducing the overall cost of access to space. Payload costs need to be reduced as well. Furthermore, there are subtle interactions between payload and launch costs. As launch costs increase, so do payload costs. To reduce the risk of on-orbit failure and the probability of relaunch, some payload subsystems are made triple redundant, increasing the cost and weight of the satellite. If launch costs can be reduced significantly, it may no longer be necessary to design to such high levels of redundancy. In addition, TAVs may make it possible to recover payloads in orbit, and if payloads were designed modularly, they could be quickly repaired on-orbit. Such payloads could cost considerably less than existing satellites. TAVs might enable a new era of low-cost access to space.

Acknowledgments

The authors thank the workshop participants and briefers for sharing their insights with RAND and with the Air Force. In particular, we thank Dr. Raymond Chase of ANSER for numerous insightful conversations and Bruno Augenstein of RAND for his expert review of this document. We also thank Maj. Ted Warnock, who at the time was an Air Force Fellow at RAND. Ted suggested that we hold a workshop on TAVs and helped to organize it. We thank Lt. Col. Jess Sponable, Maj. Mitchell Clapp (retired), and Kenneth Hampsten of Phillips Laboratories for sharing their work on transatmospheric vehicles and related subjects with RAND.

Finally, we thank Flora Grinage for her expert assistance in the preparation of this report.

List of Symbols

AF/SMC	Air Force Space and Missiles Center
ASATs	Antisatellite Weapons
CFD	Computational Fluid Dynamics
CINCs	Commander in Chief
CONUS	Continental United States
DSB	Defense Science Board
EELV	Evolved Expendable Launch Vehicle
ELV	Expendable Launch Vehicle
EMD	Engineering Manufacture and Development
FFRDC	Federally Funded Research and Development Center
GLOW	Gross Lift-Off Weight
GTO	Geostationary Transfer Orbit
HTHL	Horizontal Take-off and Horizontal Landing
L/D	Lift-to-drag ratio
LCC	Life Cycle Cost
LEO	Low Earth Orbit
LH2	Liquid Hydrogen
LMSW	Lockheed Martin Skunkworks
LOX	Liquid Oxygen
MLV	Medium Launch Vehicle
MMC	Metal Matrix Composites
NASA	National Aeronautics and Space Administration
NASP	National Aerospace Plane
NG	Northrop Grumman
OSC	Orbital Science Corporation
OSTP	Office of Science Technology and Policy
POST	A NASA Trajectory Analysis Program
RASV	Reusable Aerodynamic Space Vehicle
RCC	Reinforced Carbon Carbon
RFP	Request For Proposal
RLV	Reusable Launch Vehicle
RTD&E	Research, Development, Test and Evaluation
SLV	Small Launch Vehicle
SSME	Space Shuttle Main Engine
SSTO	Single Stage To Orbit
SSTS	Space Shuttle Transportation System
TAV	Transatmospheric Vehicle
TBMs	Tactical Ballistic Missiles
TPS	Thermal Protection Systems
TRD	Technical Requirements Document
TSTO	Two Stage To Orbit

T/W	Thrust-to-weight ratio
VTHL	Vertical Take-off and Horizontal Landing
VTVL	Vertical Take-off and Vertical Landing

1. Introduction

Transatmospheric vehicles (TAVs), with aerodynamic properties similar to conventional aircraft, have the promise of flying into space and delivering payloads into low earth orbit (LEO). A key distinguishing feature between TAVs and aircraft is propulsion. Jet engines or pure air-breathing propulsion systems cannot provide the thrust levels needed to ascend into space. Traditional rockets overcome this limitation by carrying their own oxidizer. However, traditional rocket boosters are expended after use. TAVs would be Reusable Launch Vehicles (RLVs) capable of returning to earth and flying again after refurbishment and refueling.[1] Because of their reusability, TAVs could potentially launch payloads into orbit or into suborbital trajectories at much lower costs than expendable launch vehicles.

Purpose of the Workshop

The Air Force asked RAND to examine the utility, feasibility, and cost of procuring a TAV capable of carrying out military missions. As part of this study, a workshop was held at RAND on April 18 and 19, 1995, to examine TAV technical feasibility and design issues. Experts from the Department of Defense, NASA, and industry participated (a list of workshop participants can be found in Appendix B). A variety of commercial RLV and military TAV design concepts were presented and discussed, including designs submitted for the NASA X-33 and X-34 program competitions. The proceedings of this workshop and certain ancillary analyses are summarized in this report.

Lessons from Recent Programs

In past decades a number of military TAV concepts were proposed, but none resulted in an operational system. The first National Aerospace Plane (NASP) program, which took place from the late 1950s to the early 1960s, demonstrated a number of important technologies, such as real-time air liquefaction, and operations such as hypersonic refueling, which could be important in Two Stage

---

[1]Although some use the terms TAV and RLV interchangeably, we distinguish between the two in this report. We will refer to civil or commercial reusable launch systems as RLVs, and to reusable launch systems that are designed for military-specific missions as TAVs. TAVs have more recently been referred to as military spaceplanes.

To Orbit (TSTO) vehicles. The second NASP program, during the 1980s, was equally ambitious. The 1980s NASP design called for a Single Stage To Orbit (SSTO) fully reusable system based on a combined cycle air-breathing propulsion system concept.[2] Because of the high technological risk of combined cycle propulsion and other aspects of the NASP design, this program never proceeded beyond the technology development phase. The program was canceled after an expenditure of $1.73B when it became clear the cost of an operational prototype would on the order of $10B. In light of the difficulties encountered in the NASP program, caused in part by reliance on risky propulsion technologies, only TAVs based on traditional jet engines or rocket propulsion are considered in this report. Indeed, the focus of the RAND TAV workshop was to explore the technical and cost issues associated with "conventionally" powered TAVs.

Because of the problems encountered in the NASP program, we believe it is also useful to determine whether there are any other technology "show stoppers" that could seriously impede development of a TAV even if existing or near-term conventional rocket propulsion is used and whether an X-vehicle or TAV prototype could be developed with relatively modest funding. At the workshop, several additional technology risk areas were identified. And depending upon the type of TAV design chosen for development, these technology risk areas could be more or less severe barriers to overcome. Achieving operationally useful and economical payload delivery capabilities regardless of the TAV design approach chosen will be challenging, but is certainly possible given the advances made in key technology areas in the past few decades.[3] A necessary step in determining the technical feasibility of one challenging TAV design concept has already been taken by NASA with initiation of the X-33 program. However, it is important to note that there are several other promising TAV design concepts, and some of these may be better suited for military missions and may not have all the technical risks associated with the SSTO X-33 design.

At the RAND TAV workshop, other TAV design concepts, including an air-launched TSTO TAV, were identified that may be better suited for military missions than the TAV designs submitted in the NASA X-33 SSTO competition.

---

[2]The NASP propulsion system was designed to operate like an air-breathing engine at low Mach numbers during the initial ascent phase and in ram and scram jet modes at high Mach numbers during the last phases of the ascent into space. The promise of scram jets is that they may reduce the amount of oxidizer that needs to be carried internally by the vehicle. However, scram jet technology is significantly more complex and much different from traditional rocket and jet engines. Consequently, existing rocket or jet engine designs cannot be extrapolated to the scram jet regime, especially at high Mach numbers.

[3]The space shuttle, a partially reusable launch vehicle, is based on late 1970s technology.

In addition, RAND has determined by independent cost analysis that it may be possible to develop both an X and Y vehicle prototype for an air-launched military TAV for significantly less than it would be to develop a full scale RLV based on the current X-33 design. This difference in costs is due partly to the smaller size and smaller payload delivery capability of an air-launched military TAV, but also to the fact that the first stage for this type of system would be composed of modified commercial off-the-shelf large transport aircraft, such as a Boeing 747. This cost analysis is reported in a companion RAND report.[4] The results of this analysis are summarized in Section 3.

Before any decision is made to develop a military TAV, an examination of potential TAV military missions is needed to assess their operational utility, as is a cost-effectiveness comparison of TAVs relative to other terrestrial, airborne, or space system alternatives that might accomplish the same missions.

Report Outline

Section 2 reviews the launch needs of civil, commercial, and military users and how these needs could potentially be met by new types of RLVs or TAVs. We review the economic motivations of commercial users, the size and weight characteristics of space payloads of the various user communities, potential changes to the types of payloads the DoD may deploy in the coming decades, and new or existing DoD missions that could be carried out by TAVs. We also contrast the cost and performance of a few representative TAV designs. Finally, we review the guidelines of current U.S. National Space Launch policy and suggest ways the policy could be improved while still maintaining effective coordination of NASA and DoD space launch research and development activities.

In Section 3, the TAV design options presented at the RAND TAV workshop are reviewed and issues associated with them are discussed. Technology, development, and operational employment risks (risks associated with particular launch or recovery modes or in flight maneuvers) associated with each design option are assessed based on workshop discussions or on subsequent system performance estimates done at RAND.

Section 4 discusses the technical challenges associated with development of commercial RLVs and military TAVs in general, and those associated with particular design alternatives. We examine the state of the art in key technology

---

[4]See *Life Cycle Cost Assessments for Military Transatmospheric Vehicles,* MR-893-AF, 1997.

areas, such as thermal protection systems, propulsion, and composite materials, to assess what further advances may be required to develop a commercial RLV or a military TAV.

Caveats

We caution the reader that not all the RLV or TAV concepts discussed in this report have received the same level of analysis, development, or criticism (due or undue, as the case may be). We could have chosen to drop certain TAV concepts presented at the workshop from this report for any of the above reasons. However, we have chosen not to do so, as many "immature" concepts appear quite promising and deserve further investigation. In some cases, however, certain TAV performance claims, such as LEO payload delivery capabilities, were made by the presenters. Discussions at the workshop revealed that some of these claims were considered controversial within the expert community. Rather than publish these claims without comment, we attempted to verify or disprove certain payload delivery claims of some of the less-mature TAV concepts. These analyses unfortunately took time, but have been completed and are included in Section 3. Publication of this report was delayed while the calculations were performed.

Most of the data presented in this report originate from material presented at the workshop. Consequently, the information cutoff date for that material is April 1995. The major exceptions to this cutoff date are the descriptions of the X-33 RLV concepts and their associated subsystems. The X-33 designs of the three competitors were still in flux when the workshop was held, so we updated the descriptions of these system concepts with publicly available or nonproprietary information made available in the months preceding the selection of the X-33 winning bid.

A number of proprietary contractor presentations were given at the workshop. To enable as broad a distribution as possible of this report, we kept the report nonproprietary. Similarly, some presentations were considered to contain sensitive information regarding potential future DoD space missions and means to accomplish them. These presentations or information resulting from them are not included in this report.

2. TAV and RLV Needs, Costs, and National Launch Policy

In this section, we examine the space launch needs of U.S. military, civil, and commercial space sectors. The potential development and operating costs for a small fleet of military TAVs are reviewed and compared with the costs associated with a small expendable launch vehicle. The potential development costs for commercial RLVs are also briefly reviewed. Next, we examine U.S. launch policy and the current division of responsibility between NASA and DoD regarding development of expendable launch vehicles (ELVs), commercial RLVs, and military TAVs. Finally, the implications for the DoD Evolved ELV (EELV) program of developing a commercial RLV within the current framework of the NASA X-33 program are discussed.

Motivations for RLVS and TAVs

A number of arguments have been made to justify development of military TAVs and commercial RLVs. The current fleet of U.S. expendable launch vehicles has serious limitations, in terms of both cost and performance. Although these vehicles have been improved over the years, they originate from designs for intercontinental ballistic missiles that are now nearly half a century old. These expendable launch vehicles are costly, because the entire rocket booster is destroyed after each successful launch, which also can make it difficult to determine how well equipment worked during launch or how it could be improved.[1]

Military TAVs are envisioned as reusable systems. A TAV could be composed of multiple stages or be an SSTO system. It is possible that military TAVs could eventually be operated more like aircraft than expendable rockets, implying they may be much less expensive to operate. And because the TAV could be reused a large number of times, acquisition costs could be amortized over a large number of flights instead of just one. Consequently, even though TAV acquisition costs would be higher than those for expendable rockets, the total cost for a large

---

[1]Booster rockets transmit data to mission control centers via telemetry links, but only those rocket components that are appropriately instrumented can be monitored.

number of TAV flights could be much less than an equivalent number of expendable rocket launches.

As we shall see in Section 3, there are many possible RLV and TAV design options, and these could differ significantly in acquisition cost, payload delivery capability, and perhaps in vehicle operability. Civil RLV, commercial RLV, and military TAV mission needs may also differ significantly, and acquisition agents in each sector could prefer different vehicle designs if this selection were based only on the highest-priority launch needs perceived for their own sectors.

In the next three subsections we review launch vehicle needs, launch program goals, and industry motivations that could determine future civil or commercial RLV and military TAV design concepts.

Civil Needs

NASA's space launch needs can be divided in two categories: launch of satellites for earth environmental monitoring and planetary exploration and launch of astronauts and payloads to carry out the U.S. manned space flight program (e.g., space station logistics support). To reduce costs, future NASA satellites have been reduced in size. Future planetary probes destined for the outer planets will be smaller so they can be launched by medium instead of heavy lift launch vehicles. Probes for exploring the inner planets will be reduced in size so they can be launched by small instead of medium lift launch vehicles. Consequently, a future RLV could be a cost-effective way to launch NASA satellites even if it had only a small or medium lift capability.

Space station logistics support will a require a medium to heavy lift launch capability because of the large volume of consumables needed to support space station operations. But most important from a financial point of view, sometime after 2005 serious consideration will have to be given to replacing the space shuttle with a follow-on man-rated vehicle—if the U.S. manned space flight program is to continue. Although it is possible for a shuttle follow-on vehicle to be smaller than the current shuttle, such a vehicle may fall into the heavy lift class, especially if it were also designed for space station logistics support or to support potential space station expansion or refurbishment plans. In either case, NASA will be increasingly motivated to develop a shuttle follow-on that has significantly lower operating costs and that is completely reusable, because the cost of operating the shuttle fleet takes up a substantial portion of the present NASA budget.[2]

Although estimates differ on how much a single shuttle launch costs, the number is on the order of $500M. With an average annual launch rate of eight missions a year, the annual operations cost of the shuttle program is on the order of $4B. Rockwell, the shuttle program prime contractor, has broken out these costs according to the costs of reusable and expendable or remanufacturable components.[3] Expendable or remanufacturable components consume about one third of the operations budget. Launch and landing operations consume about one fifth of the budget, as does orbiter and SSME support and refurbishment. Systems integration accounts for almost a tenth of the budget, while the remainder (about 15 percent) is used for mission and crew operations. A fully reusable shuttle follow-on could potentially reduce the operations budget by one third. Rockwell has argued that a completely reusable SSTO shuttle follow-on would reduce operations costs by up to 60 percent, assuming the follow-on vehicle was unpiloted and the payload capability of the follow-on was one-half that of the current shuttle system.

However, an RLV developed to support NASA's existing core launch needs and to replace the shuttle may have to be a heavy lift launch vehicle to support the U.S. manned space flight program as it is planned today—a program centered around the space station.

Commercial Needs

Over a decade ago, U.S. companies dominated the international space launch market. Today, Arianespace, with its line of Ariane 4 launchers and the new Ariane 5 launch vehicle, controls over 40 percent of the international launch vehicle market. In addition, Chinese and Russian launch service providers have aggressively bid for launch contracts in the international market. Launch contract restrictions on the price and number of launches the Russians and Chinese can provide have been agreed to by all major launch providers to prevent serious market disruption. However, these cartel-like arrangements cannot stay in place indefinitely, especially as Russian and Chinese economies

---

[2]The Shuttle Space Transportation System is composed of three major elements: the orbiter, two solid rocket motors, and the external fuel tank. The external fuel tank is expended during flight and disintegrates upon reentry. The solid rocket motors are recovered at sea and remanufactured. The orbiter is refurbished after each mission and the three Space Shuttle Main Engines (SSMEs) are replaced.

[3]T. J. Healy, Jr., *A Perspective on SSTO Systems*, Rockwell International, Space Systems Division, Downey, California, undated.

8

are integrated into world markets, because they represent unfair trade barriers for these countries.

As illustrated in Figure 2.1, U.S. launch vehicles either cost more or have smaller payload delivery capability to achieve geostationary transfer orbit (GTO) than those of foreign competitors. The figure displays launch costs and payload capabilities for U.S. and European launch vehicles, and indicates pricing trends observed for Russian and Chinese launch vehicles. The challenge faced by U.S. industry is to meet the $4,000 per-lb-to-orbit costs currently achieved by Russian and Chinese launch vehicles.[4]

Fortunately for U.S. launch vehicle providers, the commercial satellite launch market is booming today, primarily as a result of dramatic growth in satellite communications markets around the world. All major launch vehicle providers have substantial order backlogs, except possibly for the Chinese, who have suffered a series of recent launch failures. However, once market restrictions are lifted on Russian and Chinese rocket launches, U.S. launch vehicle providers may come under severe competitive pressures and in the long term may have difficulty surviving if they cannot upgrade vehicle performance and cut costs significantly.

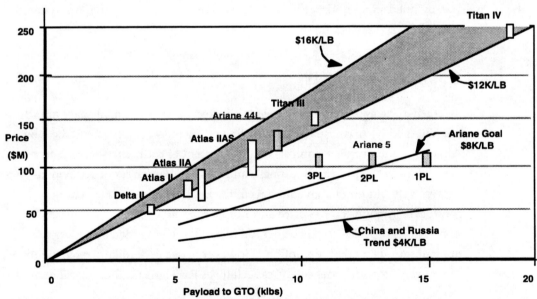

SOURCE: *DoD Space Launch Modernization Plan,* Briefing to National Security Industrial Association, Lt Gen Thomas S. Moorman, 8 June 1994.

Figure 2.1—Space Launch Costs per Pound

---

[4]The base year dollars for the "price" are not identified. The price is assumed to be in constant fiscal 1994 dollars based on the date the chart was created.

Both McDonnell Douglas and Lockheed Martin (makers respectively of the Delta and Atlas rockets) have initiated upgrade programs to increase the payload delivery capabilities of their rockets. However, it will be difficult to match the Russians and Chinese in price. New U.S. launch vehicles would have to cost about a third of what current U.S. launch vehicles cost today. Furthermore, both Russia and China subsidize their launch vehicle providers and both countries wish to earn hard currency by selling launch services to the West. Many in the U.S. launch industry have argued that a radical new approach is needed to "leap frog" foreign competition in the international space launch market. The long-term approach to this competitive threat advocated by many is to develop a completely new reusable launch system that can launch medium-sized commercial payloads weighing from 5 to 15 klb to GTO at a cost of $4,000 per lb or less. As a consequence, U.S. launch vehicle providers have collaborated with NASA in the X-33 and X-34 programs and have agreed to invest their own corporate resources in these programs.

Military Needs

There are a number of reasons why TAVs could be particularly useful for military operations. Although some of these motivations stem from shortcomings with the current fleet of U.S. launch vehicles, military motivations for TAVs differ from those of commercial or civilian users.

Military Space Missions

Potential TAV military missions or the provision of space-related forms of support for combat operations fall into four categories: space force enhancement, space force support, space control, and space force application. Below we provide commonly used definitions for these categories of space support.

- **Space force enhancement:** Operations conducted from space to support forces; e.g., navigation, communications, reconnaissance, surveillance, warning of ballistic missile attack, and environmental sensing.

- **Space forces support:** Sustain, surge, or reconstitute elements of military space systems or capabilities (includes spacelift and satellite command and control).

- **Space control:** Ensuring friendly use of space while denying its use to the enemy. To accomplish this, space forces must survey the space environment, protect our ability to use space, prevent adversaries from interfering with that use, and negate the ability of adversaries to exploit their own space forces.

- **Space force application:** Attacks against terrestrial, airborne, or space targets carried out by weapons operating in or through space.

Space Force Enhancement

DoD and national space systems have for some time played an essential role in supporting military commanders in the field, although it was not always apparent. During Desert Storm, it became clear how important DoD and national space systems could be in military operations. Since then, increasing the access of U.S. forces to DoD and national space-based surveillance, navigation, and communications assets has become a top priority for all the services.

Another key development is the growing capability of commercial remote sensing and communications satellites. U.S. military forces are planning to take advantage of these emerging capabilities and may rely a great deal more on commercial space assets in future conflicts, but will probably need to maintain selected national and DoD space capabilities to fill in coverage gaps or regional communications capacity shortfalls, and to provide highly protected communications or special-purpose surveillance capabilities.

Greater reliance on commercial space systems may make it possible to reduce DoD space acquisition costs. However, because of the properties of satellite orbits and the impossibility of determining years in advance where a conflict may occur, it is extremely difficult to optimize regional coverage or fill in coverage gaps by using satellites that are stored on orbit far in advance of hostilities. Optimal coverage of a limited area can be obtained if "gap filler" satellites were deployed rapidly into orbits that complemented the coverage of commercial, national, or allied systems.

Thus, for a number of reasons, including the need to reduce acquisition and launch costs for DoD and national space assets, relatively small satellites may replace some large of satellites in coming decades. This implies a consequent change in DoD launch needs and an increased emphasis on rapid launch of small to medium-sized satellites. Small Launch Vehicles (SLVs) may therefore play an increasing role in DoD space operations.

Most of the TAV concepts discussed at the RAND TAV workshop would not directly provide space force enhancement capabilities. One exception is the Rockwell REFLY, which would be launched by a Pegasus expendable rocket booster. REFLY is a reusable orbital vehicle that would contain integrated imaging or communications. REFLY would be capable of reaching LEO and of remaining on orbit for days or months. It would be designed to be deployed relatively rapidly and could provide a surge or gap filler capability. At the end of its on-station period, it would fire its upper stage rocket engine to deorbit and reenter and land horizontally. REFLY would resemble a small space shuttle in appearance and would take advantage of the known hypersonic characteristics of the shuttle, but would be capable of completely autonomous reentry and landing operations.

Space Launch Support

Today, most DoD payloads are medium-sized and require an MLV-class launch vehicle for their deployment on orbit. Many existing needs for MLV space launch support will likely persist in coming decades. These future needs could be satisfied by commercial or DoD expendable boosters, or possibly an X-33-derived RLV. However, as mentioned above, the mix of MLV and SLV payloads may change significantly in coming decades, and the DoD may have a growing need for SLV space launch support.

To quickly deploy small military communications, early warning, surveillance, or reconnaissance satellites, the United States would require a responsive and cost-effective SLV. One solution to these potential emerging needs is a military TAV. If such a TAV were available, less costly, and more capable, retrievable and reusable satellites could be developed that could be easily upgraded on the ground as technology advanced. Perhaps most important, however, with a rapid satellite deployment capability, United States Commander in Chief, Space Command (USCINCSPACE) would be able to tailor the satellites, space forces, and other assets under his command to support warfighting CINCs and component force commanders in a timely and responsive manner. These capabilities would also reduce the need to prioritize and "ration" selected space capabilities between different CINCs, government agencies, and other national users.

In this regard, it should be recalled that even with all the capabilities of the DoD and national space communities, shortages of satellite communications capacity and surveillance coverage of the area of operations were encountered by military users during Desert Storm. Even though a number of communications and other types of satellites were in storage and available for launch, and even though the United States had six months to prepare for the conflict, no satellites were launched over that six-month period because launch vehicles were not available for this purpose.

Space Control

Another mission area where a TAV could have significant military utility is space control. Increased emphasis on space control may be necessary because of the proliferation of advanced space systems and technologies, and because advanced foreign or commercial remote sensing and communications satellites may become available to potential adversaries. Adversaries may be able to use advanced space capabilities to their advantage on the battlefield. A TAV could potentially deploy space control payloads during wartime to selectively and temporarily deny enemy access to satellite systems that provide coverage of the theater of operations.[5] Recent RAND research indicates that a variety of useful space control payloads could be launched by an SLV or a military TAV.

General Merrill McPeak, the former Chief of Staff of the Air Force, identified space control as an essential Air Force mission and one requiring development of new antisatellite weapons (ASATs) to protect U.S. forces.[6] Political sensitivies, controversy over the role of such weapons in a strategic conflict, and their implications for the monitoring of strategic arms control agreements have so far prevented their deployment by the United States, even though the United States developed and tested an air-launched ASAT in the 1980s.

The role space systems had in supporting U.S. forces during the Gulf War has been noted by observers in the United States and abroad. Although maintaining the current strategic balance between Russia and the United States is of primary importance in the current national security environment, emerging threats and foreign space capabilities could pose significant future threats to U.S. forces. A key aspect to countering such threats may be timely delivery of space control payloads to target satellites in space.

Space Force Application

TAVs could potentially carry out space force application missions in addition to the military missions already mentioned, although a high-level national decision

---

[5] Further discussion of this subject is beyond the scope of this report.

[6]Neff Hudson and Andrew Lawler, "McPeak Presses for ASAT Options," *Defense News*, April 19, 1993.

to weaponize space would first be required. There are a number of reasons to consider acquiring such a capability.

The continued proliferation of tactical ballistic missiles (TBMs) may enable an adversary to accurately strike distant surface or space-based targets within minutes. Highly responsive ballistic missile defenses or counterforce capabilities may be needed to counter or deter the use of such weapons. If a highly responsive TAV were capable of attacking TBM launchers and was able to do so minutes after TAV launch, this type of weapons delivery capability could have significant deterrent value and could provide a global counterforce capability.

A second set of potential time-sensitive targets for TAV-delivered weapons would be enemy ground forces. It is not inconceivable that U.S. forces could have very little warning of a surprise attack by an adversary, as was the case in the Iraqi invasion of Kuwait. If a highly responsive TAV could delivering weapons against terrestrial targets such as armored vehicles, it could deter potential regional aggressors and help slow an enemy invasion, perhaps within minutes after the border was crossed and before allied cities or industrial facilities had been captured.

A third set of potential time-sensitive targets is hardened facilities or deeply buried bunkers that may contain weapons of mass destruction (WMD) or key command and control centers. Today there is concern that the United States lacks effective weaponry for destroying some classes of hardened or deeply buried targets and that potential adversaries are increasingly building large underground complexes that will be extremely difficult to attack using air-delivered weapons.[7] Experiments with developmental air-delivered weapons reportedly have not been successful to date. A potential advantage of a TAV or space-delivered weapon is the high kinetic energy such a weapon can impart to such a target, potentially enabling increased target penetration and weapon lethality. A TAV capable of delivering effective weapons against hardened and heavily defended targets could add an important new capability to the U.S. arsenal.

Military TAV Performance Characteristics

Phillips Laboratory has drafted a Military TAV Technical Requirements Document (TRD) for the Air Force Space and Missiles Systems Center that provides an initial set of technical performance criteria that can be used to guide

---

[7]Mark Yost, "The Underground Threat," *The Wall Street Journal*, 23 July 1996, p. 22.

contractor military TAV design efforts.[8] The TRD specifies a desired TAV payload size of 1,000 lb and provides a number of responsiveness guidelines. It states that a reusable vehicle should be ready for launch within seven calendar days under "normal conditions," and should be capable under "emergency or surge conditions" of doubling the flight rate and be ready for launch within hours. As discussed in more detail below, this type of responsiveness for a military TAV is closer to the payload delivery capability of Pegasus than to that of traditional heavy and medium lift ELVs.

The TRD also states that a TAV should have a minimum operational availability of 90 percent independent of weather conditions. Availability is related to reliability. The TRD provides specific aircraft-like guidance for vehicle reliability.

Launch Vehicle Responsiveness

We define launch vehicle responsiveness to be the time needed to prepare a new vehicle or one that has just returned from space for launch for a new mission. As noted above, an essential TAV characteristic for any of the military missions considered above is responsiveness. The TRD indicates the level of responsiveness should be on the order of hours, which is very similar to that of aircraft. In contrast, the responsiveness of current U.S. launch vehicles is measured in days. Nominal responsiveness estimates for the current fleet of U.S. launch vehicles versus their payload delivery capabilities are plotted in Figure 2.2. The data were compiled by RAND from interviews with space vehicle launch personnel at Vandenburg Air Force Base, Cape Canaveral Air Force Base, and the Kennedy Space Center.

From the figure it is apparent that medium to heavy lift U.S. launch vehicles are not at all responsive and that vehicle responsiveness improves dramatically with decreasing payload delivery capability. In other words ,it takes more time to prepare larger launch vehicles for operation. Consequently, existing U.S. launch vehicles, with the possible exception of Pegasus, cannot support the timelines required to effectively carry out the military missions examined above.

The degree of responsiveness required for military operations may be possible only if TAVs could be operated like aircraft and be put on alert like bombers. Alert status for a TAV would probably differ in some respects from that for

---

[8]*Technical Requirements Document for a Military Trans Atmospheric Vehicle (TAV)*, Advanced Spacelift Technology Program, Phillips Laboratory, Space & Missiles Directorate, February 1995.

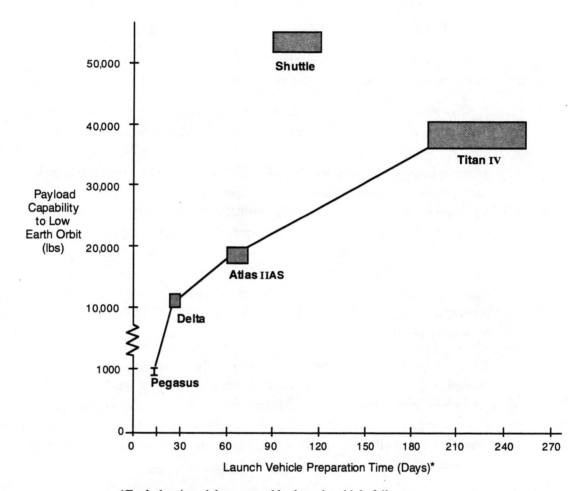

*Excludes time delays caused by launch vehicle failures

Figure 2.2—Responsiveness and Payload Capability of U.S. Launch Vehicles

ordinary aircraft if only because of the different fuels used. For example, cryogenic fuels could probably remain in the TAV for only a limited amount of time to keep the temperature of key flight systems within acceptable limits. Nevertheless, even if a TAV could fly only once every five days and if 30 minutes was needed to fuel the vehicle before launch when on alert, a TAV could still provide a much more responsive launch capability than is available from conventional launch systems today.

Aircraft-like levels of responsiveness imply aircraft-like levels of supportability and reliability. TAV checkout, fueling, and testing would have to be done routinely by standard Air Force support personnel. Faulty avionics components would be detected automatically during system self-test, and replaced by simply replacing "black boxes." This type of aircraft-like supportability should imply a much higher level of reliability than that of current launch vehicles.

Launch Vehicle Flexibility

Flexibility is defined here as the capability of a TAV to deliver payloads to a variety of orbits and to operate from a number of different bases.

The first aspect of flexibility—the ability to insert payloads into a wide variety of orbits—is a function of the launch infrastructure needed for system operation. Most current U.S. launch vehicles can be launched only from the one or two locations in the continental United States (CONUS) with an appropriate launch pad. Because of launch range safety restrictions, satellites are launched into a restricted set of orbits (defined by their orbital inclination) from each launch center. Until launch vehicles no longer expend rocket booster stages during ascent, these launch azimuth restrictions will remain in place. Pegasus, an air-launched vehicle, should not suffer from these restrictions. Pegasus can launch from ocean locations or from an isolated island like Hawaii and insert satellites into virtually any orbital inclination. An air-launched TAV could have this type of Pegasus-like flexibility.

The second aspect of launch vehicle flexibility is the ability to operate from multiple launch sites or airbases. Flexibility in TAV recovery is important not only after reentry and mission completion, but also in the event of a mission abort during ascent. Flexibility during the ascent phase would enable the TAV to have a robust set of mission abort modes and increase the failure tolerance of the system.

Launch infrastructure could affect system flexibility. If a specialized infrastructure were needed, either because of fuel storage or handling reasons or because of the size and cost of supporting gantries or launch towers, the number of TAV bases may be limited, if only for cost reasons. In addition, if only one or two bases were available, they could be valuable targets to an adversary with long-range strike capabilities. A military TAV capable of utilizing a number of different launch sites would be more survivable.

It would also be desirable for a military TAV to have a reentry cross-range capability sufficient to permit an abort back to the same or nearby launch sites in CONUS for most, if not all, possible mission profiles. This would permit CONUS basing of the TAV force and reduce or eliminate U.S. dependence on overseas bases. A large TAV reentry cross-range capability could possibly compensate for limited launch site flexibility.

Military Mission Needs Summary

Before drawing any conclusions from the above discussion of military mission needs, a thorough assessment of TAVs versus other platforms that may also be capable of performing the same missions, such as U.S. long-range bombers, is needed to determine the most cost-effective and responsive option for a broad range of targets. Such a mission analysis is beyond the scope of this report.

Another key factor in determining whether one or more military TAVs should be acquired and in what time frame is an assessment of the long-term cost benefits for such a system and the trade-off of those benefits against initial R&D and procurement costs. Ongoing RAND research indicates that a first-generation TAV, once it demonstrates reliable operations, could provide significant long-term cost savings in terms of reduced launch costs. However, acquisition of such a vehicle may be justified only if it has a multimission capability (i.e., it could support a range of missions identified above and not just a single one) and if it is based on relatively mature technologies (e.g., rocket and not air-breathing propulsion).

The mission needs identified above could result in a military TAV that is significantly different from an X-33-derived RLV designed to satisfy civil and commercial competitive market demands. As noted earlier, the X-33 contractors all envision the need for a reusable launch vehicle that can handle a payload of between 20,000 to 45,000 lb easterly to LEO. However, many of the military missions described above, and in particular the potentially important mission areas of space control and force application, may require payloads of only 1,000–5,000 lb. This size payload is closer to the capability of Pegasus than to traditional heavy and medium lift ELVs.

To be a preferred option for these potential emerging military missions, TAVs would have to be more responsive, flexible, and cost-effective than existing ELVs, long-range bombers, or other strike aircraft.

The combination of a rapid launch-on-alert capability, unpredictable launch schedule, fast turnaround time, and rapid reconfigurability to handle a variety of payloads appears to result in requirements that are uniquely military. A military vehicle capable of being launched on alert from a number of CONUS bases could be very different from a commercial RLV designed for a highly structured and predictable launch schedule that operated out of only one launch site. The demands placed on a military TAV design may have significant technical implications. For example, a military TAV may require a more robust thermal protection system and easier fuel handling than a civil or commercial RLV.

Military TAV Cost Assessment

So far we have focused on the reasons why RLVs or TAVs may be needed or desirable for commercial, civil, or military users. But how much would such systems cost and how would these costs potentially differ for RLVs or TAVs designed to serve different market niches or user needs? Below, we summarize some of the cost issues associated with developing a military TAV and present the results of a detailed cost analysis of one particular military TAV concept.

Military TAV R&D Cost Issues

Based on our assessment of civil or commercial RLV and military TAV mission needs, RAND believes there are significant potential differences between the mission needs implied or assumed in the NASA RLV program and those for a military TAV.

The differing mission and operational requirements for military TAVs or commercial RLVs could have dramatically different cost implications. Different technological solutions and types of subsystems may be needed for a military TAV. Candidate R&D tasks and cost trade-offs pertinent to developing a military TAV are described below.

One key subsystem is propulsion. A thorough examination of noncryogenic high-density fuels and high-density cryogenic tri-propellants should be undertaken. The potential operational advantages of using high-density propellants could have high payoff. A potentially inexpensive high-energy-density propellant combination is methane and liquid oxygen. The mission and operational effectiveness of various high-density fuel alternatives should be traded off against initial R&D outlays and eventual launch operations costs.

Reusable vehicle development would require trade-offs in other key technology areas. For example, manufacturing and launch operations costs of metallic or ceramic thermal protection systems (TPS) will have to be traded off against TPS performance (their durability and reliability). These factors will have an impact on TAV cross-range maneuverability and turnaround or refurbishment times. As part of the airframe and structural design process, program managers will have to analyze and test the cost-effectiveness of different TPS materials that can withstand worst-case temperatures during military flight ascent and reentry profiles.

Development of a military TAV would require an integrated design approach that will maintain strict weight margins and mass fractions, requiring close coordination of subsystem development teams.

A military TAV system-level trade analysis should be done to assess the R&D and production cost impacts of designing in the capability to operate independent of traditional launch range safety constraints. Potential increased front-end R&D cost of designing in this capability will have to be compared with possible downstream launch operations and infrastructure cost savings resulting from reduced launch delays and increased availability of the vehicle to meet mission needs.

Finally, military TAV mission abort modes would have to be analyzed to assess the technical impact and R&D and production costs for designing in the capability for safe landing at different landing sites. The X-33 and X-34 flight demonstrations will be of benefit only where the mission profiles and trajectories are similar to military TAV missions. Military TAVs may have more landing site abort options than civil or commercial RLVs and may require different mission abort demonstrator tests.

Nevertheless, many commercial RLV R&D efforts would be directly applicable to the development of a military TAV. Lightweight composite primary structures will likely be common to the designs for both types of vehicles. TPS materials are another area where significant commonality is possible. However, certain technologies and subsystems will be unique to a military TAV. For example, propulsion is likely to need unique military designs. The Air Force or the DoD would carry out the necessary R&D activities to enable future development of a military TAV and avoid redundancy with the NASA X-33 and X-34 programs.

Today, only a very limited amount of DoD funding is provided for research on military TAV R&D issues. And of course there is no DoD military TAV program.

RDT&E Cost Assessment and Comparison

A RAND cost assessment[9] for a particular military TAV concept identifies an upper-bound estimate of the total research, development, test, and evaluation (RDT&E) budget required to complete an engineering, manufacturing, and development (EMD) phase for a military TAV program, including the delivery of

---

[9]The details of this cost assessment are given in Melvin Eisman and Daniel Gonzales, *Life Cycle Cost Assessments for Military Transatmospheric Vehicles*, RAND, MR-893-AF, 1997.

subscale X-vehicle and operational prototype (Y-vehicle). The results of this analysis are briefly summarized below.

The R&D costs include nonrecurring costs to perform design trades, some of which were identified above, along with those activities required to produce a technically feasible operational vehicle design. A production phase would follow EMD with the manufacture and delivery of operational TAVs.

To gain an understanding of whether a military TAV would be affordable, we compared the dry weights, costs, and development complexity of an existing expendable launch vehicle, Pegasus, to a representative military TAV design concept, the TSTO Northrop Grumman (NG) TAV. Either vehicle is capable of potentially handling at least some of the payload range requirements envisioned for a military TAV, although the vehicles may not satisfy all military operational needs.

The results of this cost analysis are summarized in Table 2.1 for two possible variants of the NG TAV, an aerial-refueled version and an air-launched version. Both NG TAV variants are described in detail in Section 3. All costs are displayed in constant fiscal 1997 dollars. Using the TSTO NG TAV as a

Table 2.1

Military TAV R&D Affordability Assessment

Factor	Pegasus Expendable	NG RLV Aerial-Refueled	NG RLV Air-Launched
Total vehicle dry weight (lb)	6,615	25,000	34,240
Payload weight to LEO (lb)	800	1,600 – 2,000[a]	3,000 – 6,000[a]
Total engine weight (lb)		4,535 (D-57, NK-35)	3,000 (D-57)
Engine RDT&E cost[b,c] ($M)		$129.0	$87.0
Vehicle RDT&E cost[b,d] ($M)		$630.0	$590.0
Total RDT&E cost ($M)	$149.0	$759.0	$677.0
(vs. projected NASA X-33 Phase II RDT&E budget of $900M for an X-vehicle)	$71.5[e] (budget)		$350 – $600 (NG estimate)

SOURCE: Melvin Eisman and Daniel Gonzales, *Life Cycle Cost Assessments for Military Transatmospheric Vehicles*, MR-893-AF, 1997.

[a]Unmanned version.

[b]Costs based on TRANSCOST 6.0, 1995 Edition (Version 6.0) Cost Estimating Relationships (CERs).

[c]Rocket engine R&D cost based on engine weight, 200 test firings for flight certification and qualification.

[d]Vehicle R&D costs based on vehicle dry weight (excluding engines), design maturity, and team experience.

[e]Combined contractor and government budget per AIAA *International Reference Guide to Space Launch Systems*, 1991.

representative basis for a military TAV, an upper bound of $760M is obtained for the aerial-refueled TAV variant and an upper bound of $680M is obtained for the air-launched TAV variant.

As seen in Table 2.1, the estimated cost of developing and testing a military TAV X and Y vehicle is significantly greater than the $149M R&D costs for Pegasus. However, the military TAV R&D estimates of $680M and $760M, even escalated to "then year" dollars, are considerably less its the X-33 contractor estimates of total development cost for a full-scale RLV based on its X-33 designs. These estimates are given in Table 2.2 and range from $4B to $8B; they do not include the additional $1B in R&D funds that will be needed to build the X-33 subscale flight demonstration vehicle. Thus, a relatively small military TAV that costs only about $700M to develop may be a reasonable investment toward achieving the responsiveness and flexibility needed for the military missions described above.

Life Cycle Cost (LCC) Assessment

We now compare the overall life cycle costs for a military TAV and a small expendable launch vehicle. We again use Pegasus for comparison purposes. The higher front-end RDT&E budget for a military TAV should result in lower total recurring cost relative to an expendable launch vehicle like Pegasus. The recurring launch costs for an expendable Pegasus include the total production cost for each vehicle plus the fixed and variable launch operations cost. The total RDT&E cost for Pegasus of $149M can be amortized over the expected number of operational expendable vehicles procured.

As with Pegasus, the military TAV recurring cost may be comprised of fixed and variable launch operations costs. The higher nonrecurring RDT&E budget can be amortized over the total number of anticipated TAV flights. Because of reuse, the military TAV RDT&E budget can be amortized over a larger number of flights than Pegasus. Also, the recurring cost of procuring each military TAV can also be amortized over the anticipated number of launches per vehicle. The total

Table 2.2

Estimated Development Costs for Full-Scale X-33–Derived RLVs

	McDonnell Douglas	Lockheed-Martin	Rockwell International
Contractor estimated development cost	$4–7B	$4.5–5.0B	$5–8B

SOURCE: Joseph C., Anselmo, "NASA Nears X-33 Pick," *Aviation Week and Space Technology,* June 17, 1996, p. 29.

recurring cost per military TAV launch is comprised of these two amortized costs along with the total launch operations cost per vehicle.

A total LCC budget was generated for procuring one TAV and a fleet of six TAVs. To get a relative economic sense of the value of a military TAV total LCC budget, an equivalent number of launches was computed for Pegasus. LCC budgets were computed for the two variants of the NG TAV, and it was assumed that each TAV could be reused for a minimum of 100 flights and a fleet of six for 600 flights.

The results are summarized on Table 2.3. The total LCC for one operational military TAV for 100 launches is $1,867M. This same total LCC budget would only pay for only 64 Pegasus launches. The total LCC for a fleet of six air-launched TAVs and 600 TAV flights was computed to be $7,899M. The same LCC budget provides for only about 517 aerial-refueled TAV launches. And this LCC budget can provide only about 290 Pegasus launches. A more detailed discussion of the cost analysis summarized here can be found in the source cited in Table 2.3.

In the long term, an air-launched military TAV is clearly more cost-effective than Pegasus. The cost advantage for this TAV becomes even more apparent when one compares the cost to deliver a pound of payload to LEO for each of these launch systems.

Table 2.3

Military TAV Life Cycle Cost Affordability Assessment ($M)

Factor	Pegasus Expendable	NG RLV Aerial-Refueled	NG RLV Air-Launched
Total RDT&E NRE cost[a]	$149.0	$759.0	$677.0
Avg. recurring unit production cost[a]	$17.0	$60.0	$55.0
Avg. refurbishment cost/launch	N/A	$1.7	$1.5
Direct operations cost (DOC)/launch	$4.7	$7.6	$6.1
Indirect operations cost/launch	$3.1	$3.9	$3.7
Launch insurance/launch	$1.9	(included in DOC)	(Included in DOC)
Total recurring cost/launch[b]	$26.7	$13.8	$11.9
Total LCC Military TAV (1 vehicle)	$1,867.0	$1,867.0	$1,867.0
Equivalent number of launches (1 vehicle)	64	80	100
Total LCC military TAV (6 vehicles)	$7,899.0	$7,899.0	$7,899.0
Equivalent number of launches (6 vehicles)	290	517	600

SOURCE: Melvin Eisman and Daniel Gonzales, *Life Cycle Cost Assessments for Military Transatmospheric Vehicles*, MR-893-AF, 1997.

[a] All costs computed based on CERs from *TRANSCOST*, 1995 edition (version 6.0).

[b] Expendable recurring cost includes procurement cost ($17.0M) of Pegasus. Recurring cost per launch for RLV amortizes vehicle recurring cost ($60M) over 100 flights.

These costs are summarized in Figure 2.3. Because Pegasus has a maximum payload size of only 800 lb, its payload cost per pound is about $34k/lb. The payload cost per pound for the aerial-refueled NG TAV variant is estimated to be about $8.5k/lb. The payload cost per pound for the air-launched NG TAV variant is lowest of the three systems at $2.9k/lb, which is more than an order of magnitude less than the Pegasus cost, and significantly less than the cost per pound of any launch vehicle available on the market today.

National Launch Policy

The dramatic decline in U.S. market share in the international space launch market noted earlier has raised concerns over the competitiveness and health of the U.S. launch vehicle industry. As a result, the Congress in the fiscal 1994 Defense Authorization Act directed the Secretary of Defense to develop a strategic plan for the modernization of U.S. space launch capabilities. On 31 March 1994, a study team led by Lt Gen Thomas S. Moorman, Jr., produced the Space Launch Modernization Plan.

The Moorman Panel findings, while not surprising, are nevertheless troubling:

- Launch costs for U.S. government payloads continue to increase.

- U.S. demand for heavy lift ELVs has declined significantly.

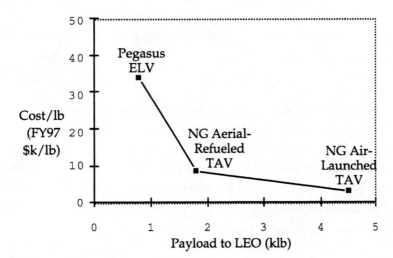

Source: MR893-AF.

Figure 2.3—Payload Costs per Pound for Pegasus and Two Military TAV Concepts

- The U.S. market has too many niches, resulting in poor efficiency and economies of scale.

- Costs of launch failures are high and increasing.

Perhaps most important, the Moorman Panel identified the need to improve the coordination of DoD and NASA space launch programs. The panel recommended a clear division of responsibility between DoD and NASA—that DoD take the lead in developing future ELVs, and that NASA have the lead role in developing RLVs.

As a result of the Moorman Panel recommendations, the Air Force initiated the EELV program under the direction of the Air Force Space and Missiles Center (AF/SMC). AF/SMC issued a request for proposals (RFP) in May 1995. The goal of the EELV program is to reduce launch costs by up to 50 percent. However, the total R&D program budget will be limited to $2B because of DoD budget constraints. It is expected that all DoD MLV-class payloads will be launched by the EELV when the new booster becomes available around the year 2000.

Depending upon the configurations eventually chosen for the EELV family of vehicles, it may eventually replace the Titan IV. Indeed, one of the strongest motivations for this program is to replace the Titan IV with a more cost-effective heavy lift launch vehicle.

Subsequent to the Moorman Panel, the Office of Science Technology and Policy (OSTP) released a new National Space Transportation Policy that gave NASA the lead for developing a new RLV. This policy's goal is to develop a commercially viable RLV that would enable U.S. industry to "leap-frog" the foreign competition.

NASA developed an RLV development plan and submitted it to OSTP for approval in November 1994. The NASA RLV program consists of a series of demonstrator programs: DC-XA, X-33, and X-34, the largest of which will be the X-33 program.[10] The Phase II selection of a single X-33 contractor, Lockheed-Martin, took place in July 1996. By the end of 1999 and upon completion of the X-33 flight demonstration program, the U.S. government and industry will decide whether and how to proceed with full-scale RLV.

---

[10]As stated by William Claybaugh, NASA Special Assistant for Commercial Programs within the Space Transportation Division, at the RAND TAV Workshop.

Corporate X-33 Program Goals

The overall goal of Lockheed-Martin is to develop a commercially viable SSTO RLV that will cost substantially less to operate than competing expendable systems. A primary corporate objective is to achieve revenue and profit targets within corporate capital investment constraints while using realistic market assessments for RLV demand. An important aspect of this corporate financial objective is to recoup investment in X-33 and full-scale RLV development costs within a reasonable time frame at an acceptable corporate investment rate of return (IRR). The business objectives that have to be satisfied to achieve these goals cover

- having the first RLV to enter the marketplace,
- building a reliable launch vehicle and achieving a first successful launch,
- meeting market-based "cost-per-lb-to-orbit" pricing targets,
- designing for low operations costs,
- establishing long-term cash flows and a predictable launch rate,
- lining up customers (e.g., anchor tenancy), and
- establishing good returns on spaceport-type launch services.

The Lockheed-Martin X-33 program is designed to achieve IRR performance measures assuming an operational RLV can capture some share of both the government and commercially forecast global launch market. According to one X-33 competitor, an operational RLV will be commercially viable only if the total development cost can be amortized over a number of projected launches and corporate tax regulations are relaxed to provide long-term tax relief on RLV capital investments.

Setting realistic market assessments will directly affect the ability to achieve corporate IRR goals. If the contractors set the cost-per-lb-to-orbit price to achieve the IRR goals within a specified payback period, and the contractors do not meet the expected number of launches per year (i.e., are too optimistic), then the contractors' available R&D funding cap may have to decrease significantly. Not projecting the market correctly may have a detrimental near-term impact on the corporate financial resources that can be invested in vehicle development. If contractors are not given anchor tenancy in the U.S. government launch market (see below), then it may not be economically feasible for them to completely absorb the financial risk to develop an RLV that can meet low operations costs and medium lift payload targets.

Several market assessments of U.S. launch vehicle demand have been made recently. The Moorman Panel envisioned between 25 and 32 U.S. government launches per year from FY2000 to FY 2010 and about another 16 U.S. commercial launches annually in the same time frame. This total demand of between 41 and 48 launches per year was divided between the NASA shuttle and small, medium, and heavy class payloads of ELVs. The X-33 competitors developed their own mission models and distributed this demand across future launch vehicles. Average demand levels varied and fell between 32 and 46 RLV launches per year. This range of launches per year is very close to the combined U.S. government and commercial launches projected by the Moorman Panel.

There was agreement among the contractors that DoD payloads of greater than 20,000 lb to low earth polar orbits that fall within the Titan-IV HLV class are outside the practical design limits for a marketable RLV SSTO concept. All of the X-33 contract competitors designed their RLVs to capture the majority of the Delta, Med-lite, and Atlas class payloads.

NASA has implemented a cooperative agreement with Lockheed-Martin to share X-33 RDT&E costs. However, no definitive agreement appears to have been reached on how development costs would be shared, if at all, for production of one or more full-scale RLVs. Statements have been made by high-level NASA officials indicating that development costs for a full-scale RLV will have to borne completely by industry because there is no money available in the NASA budget to support this activity. This view corresponds with the attitude of many in Congress, where calls for increased reliance on commercial space capabilities and markets have frequently been heard. Added to the reluctance of Congress to underwrite development of a new medium-to-heavy lift launch vehicle is a recent history of failed launch vehicle programs in which significant DoD or NASA funds were spent but no new launch vehicles were produced.

By the year 2000 or perhaps earlier, Lockheed-Martin and NASA will have to develop a comprehensive financial plan for the full-scale RLV. The financial strategy will include an economic analysis and an evaluation of commercial and civil market requirements, implementation options, and an assessment of associated financial risks. At the RAND TAV workshop, the X-33 competitors described what they believed would be essential elements of such a strategy. Common to the strategies of all the X-33 competitors was assumption of a government guarantee for a certain minimum number of RLV launches for NASA and DoD payloads. It was assumed that this guarantee, called government anchor tenancy, would be secured *before* a decision was made to proceed with development of the full-scale operational vehicle.

A second aspect of an RLV financial development plan alluded to by the X-33 competitors at the workshop was a contract termination liability agreement between the contractor and the government. Termination liability would reduce the contractor's potential financial risk in the event that (1) the government later decided to not fund its portion of the full-scale RLV development cost (if such a portion were agreed to at program start) and (2) the government later decided to not abide by the previously agreed to terms of an anchor tenancy arrangement.

A goal of all X-33 competitors at the RAND TAV workshop was that RLV would eventually become the sole source launch services provider in the U.S. government medium payload market. After an initial anchor tenancy period had expired, the RLV would compete head-to-head, on a cost and performance basis, with all remaining U.S. expendable launch vehicles, including the EELV.

Potential Conflicts in National Launch Policy

A number of possible launch system roadmaps were investigated by the Moorman Panel. One of these roadmaps is illustrated in Figure 2.4 and is the one that now appears to correspond most closely to actual events. It displays the potential overlap of the DoD EELV and NASA RLV programs.

Two key DoD decision milestones are displayed, one near term and one originally expected to be far term. The first decision concerns whether the DoD

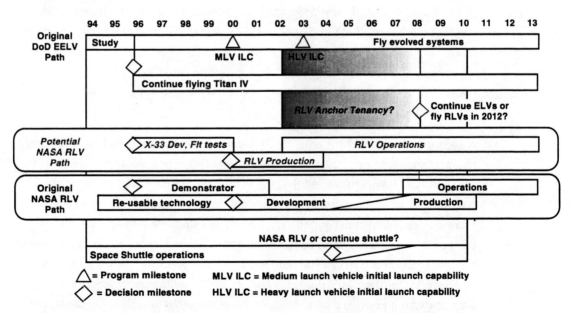

Original schedules from "Roadmap 2: Evolved ELV," pg. 44, *DoD Space Launch Modernization Plan*, Lt Gen Thomas Moorman, Briefing to the National Security Industrial Association, 8 June 1994.

Figure 2.4—DoD Space Launch Modernization Roadmap Option

will continue using the Titan IV or replace it with a heavy lift version of the EELV. This issue will be addressed when the Air Force evaluates the feasibility of proposed EELV designs that can launch both MLV and HLV payloads, and whether many if not all DoD payloads can be downsized to fit on an MLV and in what time frame.

The second DoD decision milestone illustrated in the figure concerns whether the DoD should continue using EELVs. The figure depicts this decision being made in 2008, after a NASA RLV has been developed and established an operational record. However, because of the RLV financial development plan likely to be insisted upon by industry—government anchor tenancy for an X-33 derived RLV—the decision on whether and how to proceed with the EELV program may have to made much sooner by DoD, perhaps sometime around the year 2000 or just as the EELV is scheduled to achieve initial launch capability.

As described above, NASA may pursue a cooperative development agreement to provide financial incentives to industry so industry will pay the development costs for a full-scale RLV. These incentives may include anchor tenancy at the earliest possible time in the U.S. government launch market and a guarantee that a large percentage of government payloads be launched by the RLV. Based on the presentations made by the X-33 competitors at the RAND TAV workshop, it appears that a NASA-led RLV program may be financially viable only if there is little or no government market remaining for an EELV, and vice versa.

If the NASA X-33 flight test program is successful and succeeds in generating support for development of a full-scale RLV, the EELV program may not remain financially viable in the long term. If EELV launches become too infrequent and expensive, congressional support for the program could erode. Congress, in a fiscally constrained environment, may not be willing to support both RLV and EELV programs.

A second issue raised by a possible RLV development plan based on an anchor tenancy arrangement is the possibility of repeating mistakes made with the space shuttle. Then it was promised that the shuttle would meet all the needs of DoD, commercial, and NASA users. Existing government-funded MLV programs were canceled, and consequently the Challenger disaster caused major disruptions in U.S. military, civil, and national security space programs. If some form of DoD anchor tenancy is requested for an X-33-derived RLV program, the implications for the EELV program and for DoD satellite programs need to be well understood before a national decision is made to fund RLV development.

While the promise of lower launch costs and higher reliability may be achieved by an X-33-derived RLV, the DoD and the Air Force must consider the national security implications of less favorable program outcomes. To achieve the dramatically lower launch operations costs envisioned, a high RLV launch rate may be necessary, implying that all or nearly all DoD medium-sized payloads would be launched by an RLV. An RLV failure could ground the fleet for a substantial period of time. The DoD may incur too much operational risk by relying on a small fleet of identical RLVs.

It should be noted, however, that if the DoD were to develop and deploy small "gap filler" satellites that could be launched by SLVs or TAVs, it could reduce the national security risks associated with relying on a single MLV, whether it be an RLV or the EELV. However, at the present time the DoD is not expending significant resources on developing small satellites and there is no military TAV program.

For these reasons, it appears prudent for the DoD to proceed with EELV development and to resist attempts, if they should occur, to transfer DoD payloads from the EELV to an X-33-derived RLV.

One final national launch policy issue concerns whether the DoD should be permitted to develop a military TAV, and whether potential emerging military needs warrant the start of a military program today. In considering this question, we believe it is useful to distinguish between large RLVs and small military TAVs. Certainly it would make little sense for the DoD to pursue development of a TAV with a large payload capability simultaneous with the ongoing NASA X-33 program. As indicated in Figure 2.2 and if history is any guide, it would be exceedingly difficult to develop a large RLV capable of satisfying military responsiveness needs. It is also evident that a large X-33-derived RLV could satisfy many existing DoD peacetime space lift requirements and perhaps do so in a cost effective manner.

However, potential emerging military needs may not be cost-effectively satisfied by using the current fleet of existing U.S. launch vehicles or by an X-33-derived RLV, mainly because of the launch vehicle responsiveness or flexibility needed carry out potential emerging space control and force application missions.

Because of the potentially significant design differences between commercial RLVs and military TAVs and the need for a higher-level responsiveness and flexibility for a military system, we believe the DoD should consider recommending changes to existing national launch policy to permit the DoD to vigorously study, and if necessary develop, small payload class TAVs that can meet potentially important emerging military needs.

3. Design Options and Issues

In this section we review design issues relating to future TAV development, including the advantages and disadvantages of alternative TAV launch and landing modes and those of multiple or single-stage TAV concepts. We also review the RLV and TAV presented at the RAND TAV workshop.

Launch and Landing Modes

Reusable launch vehicles or TAVs can be placed in three categories according to the modes of launch and recovery they employ. In contrast, traditional expendable space launch systems are vertical take-off systems, which by definition have no recovery modes.[1] The three categories are discussed below.

Vertical Take-off and Horizontal Landing (VTHL)

The Space Shuttle Transportation System (SSTS) is the archetypical example in this category. The SSTS first-stage elements—the solid rocket boosters and the external fuel tank—are expended about 100 seconds into launch after a vertical ascent from the launch pad. The shuttle itself continues on to orbit and after reentry lands horizontally like a airplane. Another example is the Rockwell X-33 concept, which will be discussed in more detail later in this section.

VTHL vehicles are typically aerodynamically stable in flight on their return descent trajectories, although they may, like the shuttle, have relatively low lift-to-drag ratios (L/D), which imply high landing speeds. These types of vehicles need have landing gear designed for only landing loads and not for the full vehicle Gross Lift-Off Weight (GLOW).

On the launch pad and during the early stages of ascent, the vehicle structure must be designed to take full gravity and main engine thrust loads in the vertical direction.

---

[1]The solid rocket boosters of the Space Shuttle Transportation System are recovered from the ocean after splash down, and Boeing has worked for several years on a partially recoverable first-stage booster rocket system in which high-cost engines and turbomachinery would be recovered after splash down.

Vertical Take-off and Vertical Landing (VTVL)

To date, no operational reusable launch vehicles have been developed that fall into this category. However, the McDonnell Douglas X-33 concept and DC-X flight demonstration vehicles are VTVL designs. These vehicles have ballistic missile aerodynamic characteristics and no wing structures, providing an advantage during ascent because there are no parasitic drag losses due to wings. However, this type of vehicle design can result in high reentry speeds and high aeroshell heating rates during reentry. This may lead to the disadvantage of greater thermal protection requirements on reentry and increased vehicle mass for the vehicle thermal protection system (TPS).

Landing is accomplished by restarting and firing the main engines. This increases total mission propellant requirements, but results in reduced structural weight because wings and related structures are not needed. An increase of approximately 1000 ft/sec in ideal velocity is needed for vertical powered landing.[2] Studies have indicated that there is no overwhelming advantage or difference in overall vehicle weight (GLOW) between vehicles using horizontal and vertical landing modes. However, there are increased risks of mission failure with vertical landing systems because of requirements for main engine restart, the high thrust levels potentially needed, and precise thrust vector control needed at landing and after reentry and exposure to the space environment.

Horizontal Take-off and Horizontal Landing (HTHL)

There are no current examples of an HTHL system. The Pegasus winged booster rocket is a horizontal take-off vehicle that is released at altitude from a first-stage carrier aircraft. The system is composed of a B-52 or L-1011 carrier aircraft and a winged rocket vehicle with three stages. About 5 seconds after Pegasus is dropped from the carrier aircraft, the first-stage solid rocket motor ignites. The rocket accelerates and uses aerodynamic forces to change its trajectory and pitch upwards. One advantage of an HTHL system is that lift forces can be used to adjust the ascent trajectory as needed in the atmosphere and to counteract gravity losses.

At take-off, the HTHL vehicle must possess landing gear capable of handling the full gravity loads of a fully fueled vehicle. Thus, the landing gear can be quite heavy, which has led to HTHL designs in which the vehicle first stage is a rocket or jet powered sled containing the landing gear. Once take-off speed is

---

[2]R.L. Chase, *A Comparison of Horizontal and Vertical Launch Modes for Earth-to-Orbit NASP-Derived Vehicles*, AIAA 91-2388, AIAA/SAE/ASME 27th Joint Propulsion Conference, June 24–26, 1991.

established, the second stage HTHL vehicle would separate from the supporting sled and take off like a conventional aircraft. Such HTHL systems may suffer from a significant operational disadvantage because they have to operate from air bases with extraordinarily long runways to accommodate sufficient stopping distance for the first-stage sled.

Vehicle Staging

To date all operational space launch vehicles have been multistage systems in which booster rockets separate from the launch vehicle at some point in the ascent trajectory. Because heavy first-stage rocket engines and tanks are expended during ascent, the mass of upper stages can be reduced considerably relative to the payload carried. The ratio of payload to total stage mass is considerably higher for an upper stage. In other words, vehicle staging can significantly reduce the delta-V required for the final upper stage to reach orbit. Vehicle staging may be accomplished by using a launch platform, by in-flight propellant transfer to the orbital vehicle, or by use of conventional upper stages.[3]

The launch platform can be either an aircraft or a sled, and the aircraft launch platform could carry and release the orbital vehicle in a variety of configurations. It could carry the orbital vehicle underneath its fuselage and release the vehicle in an air-drop maneuver. The orbital vehicle could be mounted on top of its fuselage and be released when in a dive or pitch-up maneuver. Or it could tow the orbital vehicle to the release altitude and launch it by releasing the tow line.

Adding stages to a launch system increases performance and the payload delivered to orbit, but vehicle complexity is increased. Each stage requires its own separate propulsion system and tankage. Stages have to be programmed or commanded to separate at appropriate times during ascent, which may require independent avionics systems for each stage, communications relays between stages, and explosive bolts or other mechanisms to ensure proper separation.

Single Stage To Orbit Systems

An SSTO vehicle would be a single integrated vehicle that would not expend components during its ascent to orbit. Such a vehicle would also reenter and land either horizontally or vertically for subsequent launch and reuse.

---

[3]Gregory, Bawles, and Ardeura, *Two Stage to Orbit Airbreathing and Rocket System for Low Risk, Affordable Access to Space*, NASA, April 1994; and U. Mehta, *Air-Breathing Aerospace Plane Development Essentials: Hypersonic Propulsion Flight Tests*, NASA TM-108857, November 1994.

Developing and demonstrating an SSTO system will be a difficult challenge because of the delta-V and vehicle mass fraction required. However, these daunting challenges may possibly be met by using advanced lightweight composite materials to reduce vehicle empty weight, high specific impulse propulsion systems to increase performance, or air-breathing engines to reduce the amount of oxidizer (and thus GLOW) required to achieve orbit. [4]

Various SSTO programs have been embarked upon in the recent past, perhaps the most notable being the NASP program, which was based on a complex air-breathing propulsion concept. The technology challenges associated with air-breathing propulsion systems and other aspects of this design approach proved so difficult that no prototype vehicle was ever built.

More recently, NASA has initiated the X-33 program, whose goal is to demonstrate key SSTO technologies by the year 2000, leading the way for an eventual operational vehicle that could replace the space shuttle and existing expendable rocket boosters. The competing X-33 designs and the winning system are described in more detail later in this section.

Operability may be one advantage of an SSTO system over multiple-stage vehicles. The latter may require additional support infrastructure because of the complexity of multiple-stage systems. On the other hand, an SSTO system may be inherently more complex than a staged system because of the additional performance demanded of the propulsion system and because of other technologies necessary to gain the performance levels needed to reach orbit.

The supporting infrastructure for an SSTO system may be smaller and less expensive than for a multiple-stage system, but this will probably be sensitive to whether a horizontal or vertical take-off mode is adopted, as this difference can distinguish between aircraft-like operations and the need for specialized space launch complex support.

Two Stage to Orbit (TSTO) Systems

The simplest multistage space launch system would have only two stages. For a reusable TSTO system, both stages would be reusable. If one imagines what a reusable TSTO system could look like, the original German Sanger HTHL concept immediately comes to mind. The first stage would use air-breathing

---

[4]F. S. Billig, "Design and Development of Single Stage to Orbit Vehicles," *Johns Hopkins APL Technical Digest*, Vol. II, Nos. 3 and 4, July-December 1990.

propulsion and operate much like an aircraft. The second stage would be a rocket-powered orbital vehicle.

TSTO Air-Launched Concepts. The German Sanger concept is but one example of a TSTO air-launched system. In the original Sanger proposal, the orbital vehicle was carried on top of a specially designed first-stage supersonic Mach 6 aircraft that had no central rear tail structure, making vehicle release relatively straightforward.

If both the first- and second-stage vehicles were designed specifically for a TSTO system, they could be integrated into a combined vehicle configuration in a number of ways. The staging maneuver could potentially be performed at subsonic or supersonic speeds. An air-drop stage separation maneuver is relatively easy at subsonic speeds, as illustrated today by Pegasus. Air launch of the orbital vehicle from on top of the carrier aircraft may be a more difficult maneuver to accomplish if the carrier aircraft is not specially designed for such a maneuver. However, it is important to note that the shuttle was successfully air launched from on top of a specially modified B-747 ferry vehicle during landing tests. The carrier vehicle used in those tests is the current Shuttle Carrier Aircraft (SCA), a modified B-747-100 with an augmented vertical tail for increased stability when mated to the shuttle. The SCA can ferry vehicles that weigh up to 236,000 lb.

Supersonic vehicle separation is also feasible and was demonstrated several decades ago in operations in which the SR-71 air-launched a ramjet-powered drone at Mach 3 speeds. The cause of the one vehicle separation failure during these SR-71 drone operations was later discovered, and it was determined that the SR-71 air-launch maneuver could be safely executed at Mach 3.[5]

An important issue for all proposed space launch systems is development cost. In the case of an SSTO system, cost may not be minimized significantly by using existing vehicle systems or subsystems. However, it may be possible to use existing aircraft for the first stage of an air-launched HTHL system. The overall acquisition cost for a TSTO system would be significantly reduced if a commercial jumbo jet were modified for this purpose (development of a new jumbo jet can cost as much as $5B, or as much as a new launch vehicle). In contrast, if jumbo jet aircraft were bought off of a commercial production line, the unit cost would probably be less than $200M.

---

[5]Private communication from Bruno Augenstein of RAND.

Potential carrier aircraft include the current SCA, the B-747-100, the commercially available B-747-400, the potential future commercial variant of this four-engine jumbo jet (the B-747-600X), and the Russian AN-224 large transport aircraft. The maximum take-off weights of these aircraft are given in Table 3.1. From the table it is apparent that planned future aircraft could provide 30 percent or more lift capacity than the current SCA.

Air-launch platform designs offer other potential advantages, such as not having to use fixed launch pads, and they could enable a dramatic departure from complex vertical vehicle integration and launch facilities. First-stage launch aircraft could operate above cloud level, which would permit bad weather to be avoided, increasing launch availability and permitting operation at altitudes where dynamic pressures during launch would be significantly reduced.

Nevertheless, special facilities at launch sites may be needed for TSTO HTHL systems, such as cranes, gantries, and support structures.

Aircraft lift performance must satisfy required system launch conditions for speed and altitude. One drawback of TSTO air-launched systems is that the size of the orbital vehicle is limited by the lift capability of the carrier aircraft. This in turn ultimately limits the scalability of these designs, and prohibits evolution to very large designs and payload capabilities.

However, by using an aircraft as the first stage one potentially gains the greatly increased reliability and operability associated with commercial aircraft. In addition, many existing and potential military TAV missions may be accomplished without needing large or even medium-sized payloads, and could conceivably be carried out by an air-launched TAV.

A possible issue regarding military TAVs is whether military missions could be performed responsively using a TSTO vehicle. The additional complexity of integrating the orbital vehicle with the carrier aircraft results in time delays.

Aerial Propellant Transfer Concepts. In aerial propellant transfer concepts, the carrier aircraft is replaced by an entirely separate tanker aircraft. In this way, the orbital vehicle or TAV can take off from the ground horizontally with its

Table 3.1

Maximum Take-Off Weights of Potential Carrier Aircraft

Version	SCA	B-747-100	B-747-400	B-747-600X	An-224
Maximum take-off weight (lb)	710,000	735,000	875,000	1,000,000+	1,250,000+

SOURCES: Robert Ropelewski, "Boeing seeks to extend jumbo monopoly," *Interavia*, April 1996.

propellant tanks largely empty. It then approaches and hooks up to the tanker and fills its tanks. Upon completion of the refueling operation, it disengages, throttles its rocket engines to maximum thrust, and ascends to orbit.

Because the TAV would be rocket powered, additional rocket engines may have to be ignited during the aerial refueling operation, and because rocket engines typically cannot operate at low throttle settings, the refueling operation would be quite challenging and could probably not be performed by an auto-pilot or remote control system. For these reasons, this type of TAV would have to be manned.

The alignment of the refueling aircraft and TAV and the degree of engine throttleability required during aerial refueling are significant safety issues for this type of design.

Another safety issue for this design is the selection of the propellant to be used in the aerial refueling operation. In one design approach, hydrogen peroxide (90 percent concentration) and kerosene have been considered; the peroxide would be the propellant transferred from the tanker aircraft to the TAV. However, if peroxide is contaminated, it can become unstable and explode. Propellant contamination during refueling would be a significant safety issue and may make such operations very hazardous.

It has also been proposed that liquid oxygen (LOX) be transferred to the TAV in an aerial refueling operation. However, the transfer of cryogenic propellants introduces other complexities and potential hazards that require careful examination. This is a potentially high-payoff technology and should be investigated more thoroughly.

Propellant must be consumed at a significant rate during the transfer process, because a rocket engine is not as efficient as an air-breather. The transfer rate is a critical design consideration for these concepts. Refueling time must be minimized and propellant transfer rate maximized.

The NASP Program

The NASP program was conceived to develop an experimental aircraft, the X-30, to explore the entire hypersonic velocity flight range. The original program goal, to insert a manned air-breathing SSTO vehicle into low earth orbit, was never realized, although more than $1.73B was spent in this effort.[6] In 1987, the Air Force asked RAND to review the status of this program. At that time, RAND concluded that many vital technology development issues remained unresolved, even after several years of intensive research.[7] The major technology risk areas identified were computational fluid dynamics (CFD) and the integrated combined cycle propulsion system that contained air-breathing and rocket components.

The Defense Science Board (DSB) Task Force also reviewed the program in 1988 and found six critical technology areas: aerodynamics, supersonic mixing and fuel-air combustion, high temperature materials, actively cooled structures, control systems, and CFD. The DSB concluded that the development schedule for all these critical technologies was unrealistic.

At that time, both RAND and the DSB concluded that the CFD state-of-the-art could not serve as the primary NASP design tool and that this state of affairs would continue to exist for a decade or more. Integrated testing of the airframe and propulsion system also could not be performed with existing ground facilities because the upper velocity limit was Mach 10 or less. Resolution of fundamental design uncertainties for such an air-breathing system would require flight tests (the largest aerodynamic uncertainty were considered to be the transition point from laminar to turbulent flow, whose location affects engine performance, structural heating, and drag). Experimental flight data was considered essential to calibrate unvalidated CFD codes.

The NASP ascent trajectory had to be depressed in the atmosphere to ensure that its engines injected enough oxygen. This led to high aeroshell temperatures during supersonic flight, which in turn necessitated the use of advanced TPS materials and active cooling of leading edges and other surfaces. The working fluid in the NASP design would have been hydrogen, so hydrogen embrittlement was a potential problem for the active cooling channels in some of the vehicle structures that would have to operate in high temperature and pressure regimes.

The NASP combined cycle propulsion system was also risky. The engine design would have had to smoothly transition from a slow speed mode to ramjet mode, and then to a scramjet mode of operation. Major uncertainties regarding the mixing of hydrogen and air at high Mach numbers remain to be resolved and could have a significant impact on the design of such a propulsion system.

---

[6]Lt Gen Thomas S. Moorman, Jr., *DoD Space Launch Modernization Plan*, Briefing to the National Security Industrial Association (NSIA), 8 June 1994.

[7]Bruno Augenstein and Elwyn Harris, *The National Aerospace Plane (NASP): Development Issues for the Follow-On Vehicle, Executive Summary*, RAND, R-3878/1-AF, 1993, and related references.

Finally, uncertainty in subsystem characteristics and in hypersonic flight conditions meant that sophisticated new control systems would have had to be developed in parallel with the propulsion and airframe and integrated with them, adding to the complexity and technical risk in the NASP air-breathing propulsion concept.

In contrast, most of the TAVs considered at the RAND workshop were rocket-powered vehicles. Such vehicles do not suffer the severe heat loads NASP would have had to endure during ascent. None of the X-33 designs presented at the workshop required actively cooled vehicle structures or surfaces. At the RAND TAV workshop, skepticism was expressed about relying on CFD codes, except in well-understood, relatively low Mach number regimes. Fortunately, the rocket-powered TAV proposals considered at the workshop are generally in the low Mach number regime during atmospheric transit, and therefore are less subject to hypersonic design uncertainties than was NASP. And because there are no air inlets for air-breathing engines in purely rocket-powered TAVs, the hypersonics of these vehicles are generally easier to understand and predict.

SSTO Versus TSTO Designs

A central debate concerning the design and development of future launch vehicles is whether the focus of effort should be on an SSTO or a TSTO system. Traditionally, SSTO designs were considered more technically challenging because of the mass fractions required. They were also more performance sensitive and subject to substantial GLOW growth if mass fraction or specific impulse (Isp) design goals could not be met. However, many of these assessments were made assuming the use of 1960s or 1970s technologies. With the development of modern composite materials and lightweight metal alloys and TPS, the overall weight of launch vehicle structures can be reduced, perhaps by up to 35 percent.[8] In principle, modern SSTO vehicle dry weights should be substantially less than earlier designs that relied on aluminum airframes and first-generation TPS materials. Indeed, it has been claimed that 1990s technologies will reduce SSTO dry weights by a factor of two from their 1960s predecessors.[9] Thus, it has been argued that it is now possible to build an SSTO vehicle using 1990s technologies and that the technical risks and performance

---

[8]Jay P. Penn, *SSTO vs. TSTO Design Considerations—An Assessment of the Overall Performance, Design Considerations, Technologies, Costs, and Sensitivities of SSTO and TSTO Designs Using Modern Technologies*, The Aerospace Corp., Space Technology & Applications International Forum (STAIF-96), January 7-11, 1996, Albuquerque, NM.

[9]Ibid.

sensitivities of such modern designs would be much less than those of earlier designs.

However, it should be noted that the same advances in materials and TPS would also benefit the mass fraction and performance characteristics of TSTO designs. It has been estimated by Dr. Karasopoulos of Wright Labs (WL/LI) that the delta-V advantage of air-launching an orbital vehicle or TAV is somewhere between 1800-2400 fps over a ground-launched SSTO system designed to carry the same size payload. If the dry weight of an air-launched TAV can be reduced, the delta-V advantage for this type of system would be enhanced in two ways. The carrier aircraft could potentially release the TAV at a higher altitude because of its reduced weight, and the TAV would require less propellant or lower Isp to deliver the same size payload to orbit because of its improved mass fraction.

The quantitative advantages of using new materials in SSTO and TSTO designs have been estimated using vehicle sizing and performance prediction codes. These codes have been used to predict that SSTO systems will benefit much more from the use of new materials than TSTO systems.[10] However, it is not clear that these predictive codes apply with equal accuracy to SSTO and TSTO systems. In the last few decades, ground-launched SSTO designs have received a great deal more attention than air-launched TSTO systems, partly because of the focus of the NASP program.

Others have argued that ground-launched SSTOs are superior to air-launched TSTOs because (1) the technology readiness levels are higher for SSTOs; (2) air-launched TSTOs are more sensitive to performance losses; (3) ground-launched systems can be scaled up in size if necessary, while air-launched systems cannot; and (4) the design fidelity of air-launched TSTOs is generally lower than current SSTO designs.[11]

The last point is certainly true. Relatively little design work has been spent in looking at air-launched TSTO concepts. It is also true that unless completely new very large carrier aircraft are developed, air-launched TSTOs may not be able to be scaled up in size to meet less-than-predicted engine performance or unanticipated growth in vehicle dry weight. However, while it is true that some air-launched concepts may be more sensitive to performance losses, it is by no means clear that all air-launched concepts are. The air-launched TSTO concept chosen for the above referenced comparison to an SSTO design was Black Horse, which is an aerial-refueled concept and strictly speaking not an air-launched

---

[10]Ibid.

[11]Lt Col Jess Sponable, *Ground Launched SSTO TAV versus Air Launched TAV*, Phillips Laboratory, PL/VTX, 2 May 1995.

design. The above analysis was also performed using a launch vehicle sizing code that may not treat SSTO and TSTO concepts with equal accuracy and that assumed certain TSTO vehicle characteristics that may not be applicable to all air-launched TSTO designs.

If air-launched TSTO concepts do have an Achilles heel, it is their lack of scalability when existing carrier aircraft are used the first stage of the system. The lift capacity of commercial and military transport aircraft is limited and transport aircraft designs themselves are not easily scalable without incurring significant new development costs. Furthermore, it would cost several billion dollars to develop a new very large transport aircraft designed from scratch to act as the first stage for a TSTO system. On the other hand, if an air-launched TSTO system employed a TAV designed for launch from a modified commercial jumbo jet, the total development cost for the entire TSTO system could be reduced because the first stage would essentially be based on a commercial off-the-shelf product.

The probability that such a TSTO system could be developed successfully is a function of the maximum payload size intended for the vehicle (or, put another way, the TAV design margins used and the lift capacity of the carrier aircraft in the overall design). Realistic air-launched TAV designs that are based on existing technologies and commercial aircraft capabilities should contain adequate design margins for TAV engine performance and structural weights, and therefore may not be able to handle the MLV size payloads envisioned for SSTO systems. Nevertheless, development of an air-launched TSTO system that is designed for small to medium sized payloads, say up to 5000 lb to a polar orbit, may be feasible and could cost substantially less than SSTO vehicles designed to lift MLV size payloads into orbit.

Current Concepts

Table 3.2 lists most of the RLV and TAV design concepts discussed at the RAND TAV workshop. Several of these concepts are based on detailed technology and design studies, while others reflect promising but newer and less thoroughly explored concepts.

In addition to the X-33 and X-34 programs being sponsored by NASA, several TAV concepts discussed at the workshop have been under active investigation in the DoD laboratory community. Among these are the Black Horse in-flight aerial propellant transfer concept and a set of air-launched TAVs being studied at various Air Force laboratories. In addition to these, an air-launched TAV design

Table 3.2

RLV and TAV Design Concepts

Vehicle	Contractor/Lab	Staging	Payload	Propulsion	Comments	Status
X-33	Lockheed Martin	SSTO	Heavy	LOX-LH2	Lifting body, VTHL, aerospike engine	■
X-33	Rockwell	SSTO	Heavy	LOX-LH2	VTHL	■
X-33	McDonnell	SSTO	Heavy	LOX-LH2	VTVL	■
X-34	OSC	Air-drop	Small	LOX-storable	HTHL, L-1011	■
REFLY	Rockwell	Air-drop Pegasus	Very small	Noncryogenic	L-1011, B-52, reusable upper stage	■
NG TAV	Northrop-Grumman	Air-launched	Small	LOX-LH2	Boeing 747	
Black Horse	Phillips Lab	Aerial-refueled	Small	H_2O_2-Kerosene	KC-135Q tanker	
Neptune	Phillips Lab	Air-drop	Small	LOX-LH2	B-1B	☑
TAV	AMC HQ (Snead)	Air-launched	Medium	LOX-LH2	Boeing 777	☒

■ Under development (NASA)   ■ Design proposed   ■ Concept proposed

☑ Concept performance verified   ☒ Concept performance problem identified   ■ Concept performance problem identified

derived from a potential X-34 proposal by Northrop Grumman was also presented at the workshop. All are discussed below.

NASA X-33 Program

The purpose of the X-33 program is to prove the technological feasibility of an SSTO vehicle. Initially, a subscale demonstration vehicle will be developed that will serve as a technology testbed and a proof of principle for a full-scale RLV capable of achieving orbit with medium or perhaps even heavy payloads (those exceeding 20,000 lb).

As part of this effort, the following core technologies will be needed:

- Lightweight reusable cryogenic tanks
- Composite primary load bearing structures
- Advanced thermal protection systems
- Advanced propulsion
- Advanced avionics.

The X-33 is intended to demonstrate technology traceability and scalability from the subscale vehicle to a full-scale SSTO rocket. Critical design characteristics include a streamlined and efficient operations concept, flight stability and control, and demonstration of SSTO vehicle mass fraction. The NASA X-33 program may also lay the ground work for a future follow-on to the NASA space shuttle. NASA representative Bill Claybaugh, who presented an overview of NASA RLV programs at the RAND TAV workshop, stated that the intent of the NASA RLV program was not to develop a shuttle II (i.e., a replacement for the current space shuttle). Furthermore, there is no specific payload requirement for the X-33 program. The X-33 industrial partners were free to determine the payload capabilities of their experimental and follow-on RLV designs. In fact, as indicated below, all the X-33 competitors sized their full-scale RLVs for the commercial satellite launch market.

The three competing X-33 are illustrated in Figure 3.1. The vehicles are shown to scale. From left to right are the Lockheed Martin, McDonnell Douglas, and the Rockwell X-33 designs. It is apparent that the Rockwell design is the largest of the three. All three X-33 designs are based on cyrogenic LOX/LH2 rocket propulsion systems.

SOURCE: NASA Marshall Space Flight Center, Internet Web Address: http://rlv.msfc.nasa.gov

Figure 3.1—Competing X-33 Vehicle Designs

The X-33 contract was awarded to Lockheed Martin on July 4, 1996. First flight is scheduled for March 1999. Sometime after conclusion of the X-33 flight test program, NASA and the U.S. government will decide whether to proceed with development of a full-scale RLV. NASA has budgeted $941M for the program through 1999 in order to develop one demonstration vehicle. NASA will reportedly use $104M of this amount to support its own program infrastructure, while $837M will go to the contractors. Lockheed Martin, as a condition of the X-33 cooperative agreement and cost-sharing arrangement associated with the contract award, will invest $212M of its own corporate resources to develop the X-33. Lockheed Martin estimates that a fleet of two to three full-size RLVs will cost somewhere between $4.5–5 billion to build following the successful conclusion of the X-33 program.[12]

Below we review the X-33 designs proposed by the three contractors.

Lockheed Martin

The winning Lockheed Martin Skunkworks (LMSW) design is a lifting body VTHL SSTO vehicle with an integrated aerospike engine. The LMSW X-33 and full-scale RLV designs are shown in Figure 3.2. The LMSW X-33 will be a 53

---

[12]S. Dornheim, "Follow-on Plan Key to X-33 Win," *Aviation Week & Space Technology*, July 8, 1996, p. 20.

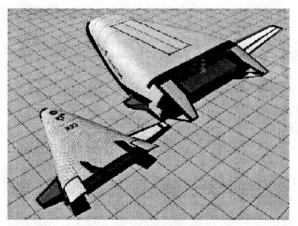

SOURCE: NASA Marshall Space Flight Center, Internet Web
Address: http://rlv.msfc.nasa.gov

Figure 3.2—Comparison of LMSW X-33 and Full-Scale RLV Designs

percent subscale vehicle relative to a full-scale RLV and will not be capable of delivering payloads to orbit. Both vehicles will employ aerospike engine designs.

Key characteristics of the LMSW X-33 and full-scale RLV are shown in Table 3.3. From the table, it is evident that even though the X-33 will be a 53 percent subscale system in terms of linear dimension, it will be much smaller in terms of volume or dry weight. The X-33 will have 12 percent of the GLOW and 31 percent of the empty weight of the full-scale system.

There are significant technical risks associated with this design, and these were identified by Dr. David Urie, the LMSW program manager, at the RAND TAV workshop. These are vehicle integration, structures, propulsion, and thermal protection. To achieve an SSTO capability, LMSW will have to achieve specific design goals in the final integrated vehicle. These include specific mass density targets for TPS surface materials, internal load bearing structures, propellant tanks, and specific impulse goals for the propulsion system.

An innovative aspect of the LMSW X-33 design is the Rockwell Rocketdyne aerospike engines planned for the vehicle. The aerospike engines will be in a linear configuration of two rows divided by a central spike. The engines will be integrated into the vehicle frame as illustrated in Figure 3.3.

Aerospike engines could have several significant advantages. They may weigh less than conventional rocket engines and their performance efficiency should not degrade as much as that of conventional engines as the vehicle increases in

Table 3.3

Key LMSW X-33 Characteristics

System Characteristic	RLV	X-33
Length	127 ft.	67 ft.
Width	128 ft.	68 ft.
Gross liftoff weight	2,186,000 lb	273,000 lb.
Propellant	LH2/LOX	LH2/LOX
Propellant weight	1,929,000 lb.	211,000 lb.
Empty weight	197,000 lb.	63,000 lb.
Main propulsion	7 RS2200 linear aerospikes	2 J-2S linear aerospikes
Liftoff thrust	3,010,000 lb.	410,000 lb.
Maximum speed	Orbital	Mach 15+
Payload (100 nmi/28.5 deg orbit)	59,000 lb.	NA
Payload bay size	15 x 45 ft.	5 x 10 ft.

SOURCE: S. Dornheim, "Follow-on Plan Key to X-33 Win," *Aviation Week and Space Technology*, 8 July 1996, p. 20.

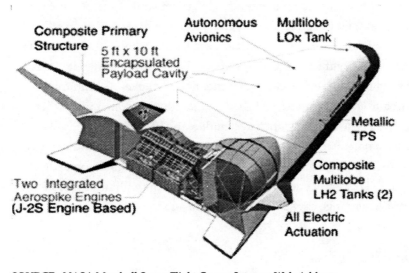

SOURCE: NASA Marshall Space Flight Center, Internet Web Address: http://rlv.msfc.nasa.gov

Figure 3.3—Features of the LMSW X-33 Design

altitude. Engine weight would be reduced because engine gimbals, mounts, actuators, and hydraulics will not be used. Instead, thrust vectoring will be accomplished by throttling different engine segments.

Another attractive feature of the full-scale RLV aerospike engine design is that it will operate at a relatively low chamber pressure of 2250 psia, which should increase engine lifetime and may reduce the need for engine refurbishment. It should be noted, however, that the aerospike engine will have to operate at 445 sec of Isp (in vacuum) in order for the LMSW X-33 to demonstrate SSTO feasibility.

A second innovative aspect of the LMSW X-33 design is the use of metallic TPS on all external surfaces except for the leading edges, where advanced carbon-carbon composites will be used. The use of metallics is made possible by the lifting body design because this body shape reduces heating loads and surface temperatures during reentry. Metallic TPS may be more durable and require less refurbishment and repair than ceramic tiles, thereby enabling low cost RLV or TAV operation and increased vehicle responsiveness.

The main vehicle structure will be composed of graphite epoxy composite except possibly for the oxygen fuel tanks, which may be made of aluminum, and the control surfaces, which will be made of titanium.

At the workshop, it was remarked that there may be major differences between a military TAV and a commercial RLV. For example, a TAV may require a horizontal take-off capability to enable it to operate out of many different airbases. And it may require a significant cross-range capability in either suborbital or orbital missions to deliver payloads quickly to their required destinations. In contrast, an RLV designed to serve the commercial launch market need not have either capability mentioned above. To minimize infrastructure costs, a commercial RLV would operate from only one launch site and may well be a vertical launch system like the LMSW X-33. It is also important to note that the lift-to-drag ratio of the LMSW X-33 lifting body design may not be not high enough (it has an L/D of 1.2 at hypersonic speeds and a maximum L/D of 4.5 at subsonic speeds) to carry out military missions where a significant cross-range capability would be needed.

McDonnell Douglas

The McDonnell Douglas X-33 entry was a VTVL SSTO design with ballistic hypersonic characteristics. McDonnell Douglas X-33 and full-scale RLV designs are illustrated in Figure 3.4. The full-scale RLV would be about as tall, at 185 ft, as the Space Shuttle on the launch pad. It would be 48.5 ft across. RLV GLOW would be about 2.4M lb and it would have a dry weight of 219,000 lb. The RLV would use eight new Rocketdyne LOX/LH2 rocket engines.[13]

---

[13]"NASA Nears X-33 Pick," *Aviation Week and Space Technology*, June 17, 1996, p. 29.

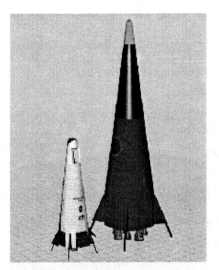

SOURCE: NASA Marshall Space Flight
Center, Internet Web Address:
http://rlv.msfc.nasa.gov

Figure 3.4—McDonnell Douglas X-33 and Full-Scale RLV Vehicles

The payload capability of the RLV would be 45,000 lb to LEO, 22,000 lb to the
space station, and 16,000 lb to geostationary transfer orbit. It would have a
payload bay size of 16.5 by 35 ft. The estimated cost to build the full-scale RLV
after successful completion of the X-33 program is $4–7B.

The primary structure would probably be made of composites as would the LH2
propellant tanks. The LOX tank would probably be composed of aluminum-
lithium alloy. One of the design issues discussed at the RAND workshop was
that if the primary structure were comprised of composites, would a very large
autoclave be needed to produce the full-scale vehicle—i.e., would the full-scale
vehicle have to fit inside of the autoclave?

The McDonnell Douglas X-33 design relies on ceramic TPS materials and most
likely employs advanced carbon-carbon composites at leading edges and on the
nose cap. This X-33 vehicle would be about a 50 percent subscale model of the
full-scale RLV. In addition, this design relies on a single Space Shuttle Main
Engine (SSME) for the main propulsion system. Key propulsion technology risk
areas identified by Dr. William Gaubatz at the workshop were the thrust to
weight ratio and throttling capability of the main engine or engines.

Dr. Gaubatz also identified significant weight uncertainties in the propulsion,
tankage, TPS, and structures areas, regardless of which design was selected in
the X-33 competition. The weight uncertainties identified in these subsystems
were 5 percent of total vehicle empty weight for propulsion, 3 percent for
tankage, 3 percent for TPS, and about 2 percent for structures. These

uncertainties will have to be reduced in the X-33 program to proceed with confidence in building a full-scale SSTO RLV.

Some other important issues discussed by Dr. Gaubatz were

- mass fraction characterization (i.e., adequate margins to account for weight uncertainties identified above),
- achieving aircraft-like operability/supportability over a 10 to 20 year vehicle lifetime,
- propulsion systems with high Isp and thrust to weight ratio and with excellent operability, enabling cost-effective number of flights between repairs and engine overhauls, and
- aerodynamic designs with sufficient cross-range, stability, and control during reentry.

McDonnell Douglas emphasized the experience base it has acquired with the DC-X program. The DC-X1 is a 1/3 scale vehicle made to demonstrate quick turnaround operations with a rocket-powered vehicle. It is not intended to validate a VTVL SSTO design. It was emphasized that DC-X was not just a vehicle demonstrator but a total system in which the aerodynamics, controls, and operations and support are demonstrated. One of the goals of the DC-X is to go from a six-day turnaround time to three days. One of the features it has to demonstrate is the ability to accommodate failures at any time during the flight envelope and still be able to return safely (i.e., without catastrophic failure).

Rockwell

This design concept is a VTHL SSTO vehicle with a composite wing and tail, aluminum/lithium (Al/Li) LOX tanks, composite LH2 tank, and an improved bad weather landing capability using durable and survivable TPS materials. The Rockwell X-33 and full-scale RLV designs are illustrated in Figure 3.5. The RLV GLOW would be about 2.2M lb and the vehicle would have a dry weight of 296,000 lb. Mass fraction goals for the vehicle are a 89.5 percent propellant mass fraction and a 2 percent payload mass fraction. The full-scale vehicle would be 213 ft long and have a wingspan of 103 ft. It is estimated by the contractor that it would cost about $5–8B to build a full-scale RLV.[14]

---

[14]Briefing presented at Rockwell X-33 RLV User Expo, Downey, California.

SOURCE: NASA Marshall Space Flight Center, Internet
Web Address: http://rlv.msfc.nasa.gov

Figure 3.5—Rockwell X-33 and Full-Scale RLV Vehicles

The RLV would be capable of placing a 43,000 lb payload in LEO and a 12,000 lb payload in geostationary transfer orbit, and it would be able to accommodate large payloads in its 45 by 15 ft payload bay. Rockwell considered both a solid and a cryogenic upper stage, but is not yet fully convinced that the latter can be carried safely in the RLV payload bay. The full-scale vehicle would also be capable of landing on a 10,000 ft runway and so could land in an emergency at a number of runways around the world.

The Rockwell X-33 design would be a 50 percent subscale vehicle capable of suborbital flight demonstration using 1 SSME and 2 RL-10-5A engines. Rockwell has decided not to use an aerospike engine because of the technical risk involved. One of the risks identified at the RAND workshop is controlled flight using aerospike engine thrust vectoring at max Q, which occurs at about 25 kft. The X-33 vehicle would be designed to take full RLV thrust loads and major portions of the vehicle, including the thrust structure, wings and LH2 tanks, would be composed of graphite epoxy composites.

The full-scale RLV concept would depend on the use of supercooled propellants. This provides a 10 percent volumetric savings with the LOX tanks and a 6–7 percent volumetric savings with the LH2 tank. This technology would be demonstrated with the SSME in the X-33 program.

Rockwell planned to use six Rocketdyne RS-2100 engines in the full-scale system, with the goal of not having to refurbish the engines (including turbopumps) for 20 flights. No cost estimates were given for engine development costs. The RS-2100 would have a vacuum Isp of 450 sec, a thrust to weight ratio of 83 to 1, and would operate at a relatively high chamber pressure of 3250 psia.

The Rockwell X-33 and RLV designs would rely on TPS blankets on all exterior surfaces except the leading edges, where high-density ceramic tiles with a density of 20 gm/cc would be used. Ceramic tiles may still have to be used on some high-impact surfaces, however. Rockwell had an operability goal of reducing the time needed for TPS refurbishment between flights by more than a factor of ten (relative to the space shuttle) to about 1500 hr.

NASA X-34 Program

The purpose of the NASA X-34 program is to provide low-cost and early opportunities to test new high-risk RLV technologies that cannot be test flown on the shuttle and that may be too risky to use in the X-33 program. Originally, the X-34 program was awarded to an industry team composed of Orbital Sciences Corporation (OSC) and Rockwell. However, because of program cost growth and differences between the industrial partners over the choice of engine, the partnership was dissolved. The original program goals included the development of a suborbital air-launched vehicle capable of reaching speeds of between Mach 12 to 14 at a peak altitude of 100 miles. The full-scale system, if developed, would then deploy payloads to orbit by using an upper stage. Another goal of the original X-34 program was to gain early RLV operations experience and to discover flight test "lessons learned" that would be useful in the X-33 program.

Orbital Sciences Corporation (OSC) X-34 Design

The OSC X-34 is composed of a hypersonic reusable rocket system and a conventional carrier aircraft. A design goal is to reduce launch costs from $12M for Pegasus to $5M for an X-34-derived vehicle. Originally, the X-34 was to be air-dropped from the L1011 or air-launched from a NASA B-747 SCA. The two original versions of the X-34 were quite different. It appears that the B-747 version may be more risky because significant wing area would be required and could impact the vehicle mass fraction.

The original X-34 development and flight test plan had the following components. Two airframes were to be built. The first airframe without propulsion system was to have undergone static load ground and captive carry tests. The second airframe was to have been test-fired at Phillips Lab on a test bench with full loadings during a simulated launch sequence using flight software. Suborbital flight tests would have then taken place to assess TPS endurance. A steep flight path angle was planned, to quickly heat the vehicle to a high temperature and thereby model reentry from orbit. The test flights were planned for late 1998 and 1999.

After the original X-34 industry team was dissolved, the X-34 contract was recompeted and awarded to OSC. The program was restructured to accommodate reduced program funding. The new vehicle will be much smaller than originally planned. It will be 58 ft long, have a wingspace of 28 ft, and a GLOW of 45,000 lb. In comparison, the original version of the X-34 had grown in GLOW to 140 klb, or a two-thirds scale shuttle.

The current version of the X-34 will be designed for 25 flights per year. The original X-34 contract was structured with NASA paying $70M of program costs, while OSC and Rockwell were to pay $50M each. For the new contract, NASA will contribute $50M and OSC an unspecified amount.[15]

Northrop Grumman (NG) X-34 Concept

Although this vehicle design concept was not formally submitted in the X-34 program competition, it is an interesting design and could have value as a TSTO air launched military TAV. This vehicle would be launched from on top of a NASA B-747 SCA and deliver a 1-6 klb payload to LEO. The B-747 launch platform would transfer LOX and LH2 fuels to the orbital vehicle.

The orbital vehicle would resemble a scaled-down space shuttle and would have its aerodynamic characteristics. It would have a GLOW of about 180,000 lb and a cross-range capability of 1100 nm. The fully loaded orbital vehicle would have a higher wing loading than an empty shuttle. Consequently, care must be taken to guarantee positive vehicle separation and to provide adequate clearance from the aircraft during the staging maneuver. The contractor has indicated that vehicle drag may be reduced relative to the shuttle by 20 percent, making this maneuver easier to execute. This reduction in drag would need to be confirmed using computational fluid dynamics.

The vehicle would use two D-57 Russian engines, which have been licensed from the Russians by Aerojet. These engines are fully throttleable and could run with a smaller nozzle (88 in. versus 143 in.) than originally designed. The two engines would produce 88 klb of thrust each. The Russian engine manufacturer has built 105 engines and Phillips Lab has performed over 53,000 seconds of engine testing. Given the performance of the D-57 engine, Northrop Grumman has

---

[15]"NASA Gives Orbital Second Shot at X-34," *Aviation Week and Space Technology*, June 17, 1996, p. 31.

estimated an orbital vehicle payload delivery capability of 1,000 to 3,500 lb to polar orbit and 3,000 to 6,000 lb to an easterly orbit. These payload weights carry no margins.

The technology risks identified by Northrop Grumman at the RAND workshop were structural weight uncertainty, TPS weight and performance, safe vehicle separation from the 747, and Aerojet capability to produce the Russian engines. The TPS materials used would be different from the materials used on the shuttle. The new materials would have an average density of .5 lb/sq ft. A major concern is further reduction in TPS weight.

Other options for this vehicle concept are to configure the orbital vehicle for a two-person crew or to develop a modified vehicle that would be capable of using high-density propellants and of executing an independent ground take-off, aerial refueling, and ascent to orbit mission profile.

Additional TAV Design Options

Several small TAVs with varying levels of technological maturity that may have military utility were proposed at the RAND TAV workshop. Further analysis and systems definition work are required to assess the feasibility of these designs and their mission utility. Some of the issues surrounding these concepts are discussed below.

Black Horse

Black Horse is an aircraft-like vehicle that would be about the size of an F-16C (see Figure 3.6). It would use H_2O_2 (peroxide) and kerosene as propellants. At GLOW, it is estimated to have a weight of 184,000 lb. This concept uses in-flight propellant transfer to provide the delta-V needed to reach orbit. Gross TAV take-off weight would be 25,000 lb. A KC-135Q tanker with isolated tanks built for the SR-71 program would off-load the bulk of the peroxide needed to achieve orbit. A major issue is whether effective flight control can be maintained during refueling. Because the lift to drag ratio of the TAV changes from 9 at hook-up to 4 at ascent, an additional engine may have to be started during the propellant transfer process.

The payload mass fraction that the Black Horse concept can achieve and the maximum payload size this design option can scale up to require further careful analysis. RAND carried out an independent analysis of Black Horse payload mass fraction capabilities using POST, a NASA trajectory analysis program, and

SOURCE: Ferrand, Kerry, Spacecraft and Technology Images, Internet Web Address: http://202.50.196.210/kk/st.html

Figure 3.6—Black Horse TAV During Fueling Operation

determined that the vehicle in its current configuration could not achieve orbit. Even if the Black Horse refueling operation could be safely executed and the vehicle could be modified to reach orbit, a potential drawback of this design may be that it will be capable of lifting only very small payloads (i.e., less than a thousand pounds) into LEO. An issue is whether very small satellites could satisfy military mission requirements.

The orbits accessible by Black Horse may also be limited. Satellite delivery to polar orbits may not be feasible, and it may not be possible to deliver satellites to equatorial orbit without significant redesign of the system. A number of options to overcome these payload limitations were suggested at the workshop: use of an upper stage, use of an air-breathing engine, or refueling ballistically (by flying two aircraft on parallel trajectories, transferring oxidizer to the orbiter, and then returning the dry aircraft). These options could possibly increase payload capability to perhaps 10,000 lb, but would introduce additional system development and complexity.

The use of kerosene and peroxide would require development of a new engine. Although this type of engine was developed and used by the British in the Black Knight project, the latter's design may not be directly applicable to current designs, such as Black Horse.

The H_2O_2-kerosene rocket engine design may have significant technical risk. An important engine performance issue is whether the chamber pressure is too high, which raises maintenance and operability concerns. A staged combustion cycle is used in which a catalyst decomposes H_2O_2 into steam and oxygen before entry into the turbopump. A concern was raised by workshop participants that a high-temperature, oxygenated environment raises serious turbopump survivability issues.

The aluminum Black Horse structure weight was independently checked by Boeing. Boeing's weight estimate is 8 percent higher than the original one, introducing another concern regarding the design feasibility.[16]

The impact of life support systems is yet another source of concern and uncertainty for this concept. Pressure suits for crew members would be required, putting a limit on how long a pilot could remain in orbit. Fatigue becomes a significant factor after 8 hours in a pressure suit, and a 24 hour mission is considered unacceptable.

Air-Launched TAV

Ken Hampsten of Phillips Laboratory presented an initial three-stage-to-orbit air-launched TAV design that would use NK-31 and D-58M Russian rocket engines. The first stage carrier aircraft would be a B-1B. A modified NK-31 engine would deliver 90,000 lb of thrust and an Isp of 355 sec using a 114 in. nozzle and would power the air-dropped vehicle's first stage. The third stage orbital vehicle would use a D-58M, which would burn LOX and kerosene and deliver 19,000 lb of thrust and an Isp of 353 sec.

This concept is designed to provide first- and second-stage mass fractions of .88 and .83 with 12,000 lb of propellant. It was indicated the orbital vehicle would have a 2,000 mile cross-range and could deliver payloads measuring up to 8 ft in diameter.

Boeing Advanced Concepts

Vince Weldon of Boeing discussed design and propulsion issues associated with TSTO air-launched TAVs.

One approach briefed is to modify a B-747 to carry $LOX/LH2$ propellants for a medium lift TSTO air-launched vehicle and LOX/CH_4 propellants for a military TAV (to take advantage of the higher density of methane). However, one

---

[16]Comments made by Boeing Co. representatives at the RAND TAV workshop.

drawback of using methane as a TAV propellant is that there are no engines currently available off-the-shelf. A second approach is to modify the B-747 with GE-90 engines on the two inboard pylons. This would provide a 53 percent increase in thrust for the first-stage carrier aircraft. Boeing has investigated using the Integrated Powerhead Demonstration Engine being developed at Phillips Lab for the second-stage TAV. Boeing estimates that an air-launched TAV using this engine could carry up to 30,000 lb to LEO at the Eastern Test Range using a LOX/LH2 propellant combination.

Finally, Boeing has investigated the feasibility of LOX in-flight transfer (it is dense and so should pump rapidly), stable separation of a fly-back wing design, and landing site needs for air-launched TAVs. If in-flight TAV LOX fueling were employed using a second tanker aircraft, air-launched TAV GLOW could be doubled from 250,000 lb to 500,000 lb.

4. Technology Challenges

A number of technologies are important for determining whether development of a TAV is feasible. Minimizing vehicle empty weight is highly desirable for any vehicle concept and critical for SSTO concepts. This will require the integration of lightweight composite materials into the vehicle airframe and subsystems. Rapid turnaround between missions, cost effective operation, and high payload mass fraction characteristics will also require development of lightweight, robust, and durable TPS materials.

Another key technology is propulsion. The vehicle's rocket engines will have to operate as efficiently as possible and provide the required delta-V to reach orbit. Rapid turnaround and low-cost operations also require that engines be durable, damage tolerant, easily inspectable, and capable of rapid and safe shutdown.

If the orbital vehicle is air launched or aerial refueled it becomes desirable (and perhaps necessary) to have highly throttleable engines and to maximize propellant density. With the use of high-density propellants, the size of the orbital vehicle and its empty weight can be minimized, permitting air launch using existing transport aircraft or horizontal launch from existing runways and aerial refueling with existing tanker aircraft.

In this section, we explore these technology issues by reviewing the four technology areas of primary importance for the development and design of TAVs: propulsion, materials and structures, thermal protection systems, and systems integration. This review is based upon discussions that took place at the RAND TAV workshop, on recent technical publications, and on material made available to RAND by aerospace contractors.

Propulsion

For propulsion systems, high efficiency (i.e., high specific impulse) and high thrust-to-weight (T/W) are primary performance goals. Cryogenic LOX-LH2 propulsion provides a specific impulse in the vicinity of 460 seconds. Higher efficiency may possibly be gained by adding high energy density materials, although the LOX-LH2 propellant combination is the most efficient one used in existing operational launch systems. An additional advantage of LOX-LH2 is that the exhaust product is water, which is environmentally attractive for routine operations.

A related consideration is the handling characteristics of TAV propellants. Cyrogenic fuels are more energetic, but introduce fuel handling and storage problems that may limit the number of launch bases a vehicle can operate from. The use of cryogenics requires appropriate logistics support to be in place, including the ability to store and process LH2 at very low temperatures. The complications involved in handling cryogenic fuels may affect TAV responsiveness and limit turnaround time even when in a quick-reaction or alert status mode. The most difficult problem concerning cryogenic propellants is considered to be handling LH2 in an operational environment. Hydrogen is difficult to contain, and leaks pose an explosion hazard. Another design complication arises from the very low boiling point of liquid hydrogen (20° K). This introduces material compatibility issues for both the vehicle LH2 tank and the pipes that deliver liquid hydrogen to turbopumps or rocket engines, as some metals lose strength and become brittle at such low temperatures.

On the other hand, past DoD research on the handling of cyrogenic fuels demonstrated quick-reaction fuel-loading times on the order of an hour to a few hours. It can be argued that a two-hour LH2 loading time may not introduce too great a time delay for TAV military missions. In addition, extensive experience has been gained in handling cryogenic propellants for expendable rocket systems.

Nevertheless, the use of LH2 presents additional design challenges for some military TAV concepts. If liquid hydrogen is used, propellant tank and overall vehicle size may be significantly larger than they would be if high-density propellants were used. It may be more difficult to design and build an air-launched LH2 powered TAV. Consequently, it is important to keep the trade space open between cryogenic and noncryogenic fuels, especially for air-launched concepts (methane and liquid oxygen may be an inexpensive propellant combination well suited for air-launched systems).

The maturity of proposed rocket engines is also an important design consideration for TAVs if up-front development costs are to be minimized.

Rocket engines can be classified as conventional bell-nozzle rocket engines or as aerospike engines. We next review their capabilities and a few representative designs.

Conventional Rocket Engines

There are a number of conventional rocket engines that are either off-the-shelf products or that could be used to propel TSTO TAVs. Several highly capable conventional bell-nozzle engines are currently in production but would require modification for use in an SSTO vehicle (specifically, T/W values for these engines would have to be increased; see below). An example of such an engine is the Space Shuttle Main Engine (SSME), which Rockwell and McDonnell Douglas proposed to use in their X-33 designs.

Highly responsive TAVs will require engines that can be maintained easily and that have high reliability. The SSME, while mature, is not designed for the routine maintenance and rapid turnaround of TAVs.

Conventional high-performance LOX/LH2 engines such as the SSME utilize a staged-combustion cycle. Figure 4.1 reveals schematically the differences in rocket combustion cycles. In a staged-combustion cycle, either the fuel or oxidizer is channeled completely through a "precombustion chamber," where it is mixed with a portion of its counterpart to run the turbopumps. The products are then channeled to the thrust chamber, where they are joined with the rest of the fuel/oxidizer counterpart to complete the combustion process.

The gas-generator cycle, which is used in the aerospike engine (see the next subsection), channels part of the fuel and oxidizer to a generator that runs the turbopumps and then expels the products of combustion. The remainder of the fuel and oxidizer is channeled straight to the main thrust chamber. In the staged-combustion cycle, therefore, all of the fuel and oxidizer are used as the propellant, whereas the gas generator cycle "wastes" some of the propellant mass by expelling it overboard.

However, since the staged-combustion cycle's precombustion chamber is essentially connected in series with the main thrust chamber, the pump pressure necessary to maintain a given thrust chamber pressure is higher than the pump pressure in a gas-generator chamber, where the precombustion chamber and the main thrust chamber are linked in parallel (see Figure 4.2). Therefore, staged-combustion engines generally exhibit higher Isp values than similar gas-generator engines (2–5 percent), but durability is lower because of higher pump pressures and reliability is lower because of increased complexity.

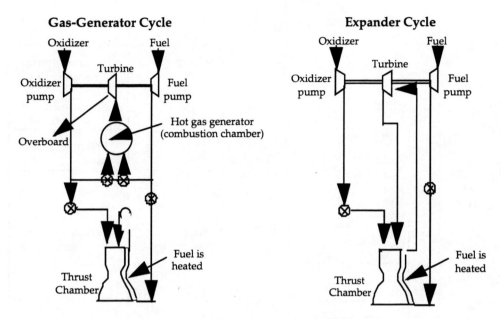

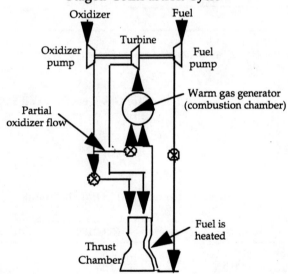

Source: R. Humble, G. Henry, and W. Larson, *Space Propulsion Analysis and Design*, McGraw Hill Inc., 1995.

Figure 4.1—Rocket Engine Combustion Cycles

The performance characteristics of two staged-combustion engines, the SSME and the Russian RD-120 engine, are listed in Table 4.1. Note the high chamber pressures in both cases.

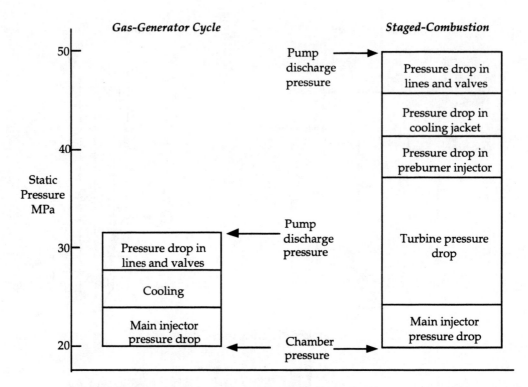

Source: R. Humble, G. Henry, and W. Larson, *Space Propulsion Analysis and Design*, McGraw Hill Inc., 1995.

Figure 4.2—Comparison of Pressure Levels for Gas-Generator and Staged-Combustion Cycles

Table 4.1

Comparative Performance of LOX/LH2 Engines

	SSME Block II (1)	RD-0120 (1)	Aerojet/Lyulka D-57
Sea-level thrust (lb)	395,000	330,000	NA
Vacuum thrust (lb)	470,000	441,000	89,600
Isp (sec) in vacuum	453	455	456.5
Chamber pressure	3200	3170	1603
Mixture ratio	6.0	6.0	5.8
Area ratio	77.5	85.7	143
T/W	51	43	48
	staged combustion	staged combustion	staged combustion
Propellants	LOX/LH2	LOX/LH2	LOX/LH2
State of development	in production	in production	in production

SOURCES: (1) National Research Council, *Reusable Launch Vehicle Technology Development and Test Program*, Washington, D.C., 1995; (2) J. D. Elvin, "X-33 Ascent and Reentry Trajectory Analysis," Lockheed Martin Skunk Works interdepartmental communication; (3) A. Wilson (ed.), *Jane's Space Directory*, 11th Edition, UK, 1995.

The SSME has proven to have excellent performance and reliability on space shuttle missions. However, it requires extensive turnaround time for inspection and refurbishment of components, and its thrust-to-weight ratio is inadequate for SSTO missions. The National Research Council reported that an SSTO RLV would require a T/W value between 75 and 80. Programs to upgrade the shuttle are projected to improve the T/W value to 58 for the Block II+ version and near 70 for the Block III version. New turbopumps, heat exchangers, valves, and combustion chamber would enable the projected performance enhancement, as well as reduce the turnaround time required for maintenance. The current SSME must be pulled between flights to replace the turbopumps. The Block II version is projected to be launched 10 times before the engine must be pulled. As it is, however, the performance of the SSME is adequate for the TSTO missions of the space shuttle. Therefore, it might not be necessary for a TAV designed for TSTO to have engines with T/W values as high as 75-80. Improvements in reliability alone may be adequate, then, to allow the SSME to meet the needs of a TSTO TAV.

The Russian RD-0120 has similar performance to the SSME. It is desired for its deep throttling capability (25-114 percent, compared to 65-104 percent for the SSME). However, it has an even lower T/W. Improvements similar to those for the SSME are planned for upgraded versions of the RD-0120.

As shown in Table 4.1, the Aerojet/Lyulka (Russia) D-57 engine, proposed for use in the Northrop Grumman TAV concept, has an Isp and a T/W that are comparable with the two larger engines discussed above. Because the D-57 is a much smaller engine, it produces less thrust, but it also weighs much less, even though it is based on late Soviet 1960s materials technology (this engine was originally developed for the Soviet lunar program).

Depending on the design of the TAV and its ascent trajectory, the high thrust capacity of engines targeted toward SSTO missions may not be necessary for TSTO missions. If this is the case, smaller, simpler propulsion systems may provide adequate performance for such vehicle designs. In his TAV proposal, Ken Hampsten of Phillips Laboratory suggested two existing LOX/kerosene engines for the propulsion system of an air-launched TAV. Although these engines have lower performance than LOX/LH2 engines, they do not necessitate the high fuel storage volumes and complex cryogenics of LH2 storage tanks and feed lines. The use of LOX/kerosene engines results in a relatively simple, easy-to-maintain propulsion system.

The performance characteristics of the two LOX/kerosene engines suggested are shown in Table 4.2.

Table 4.2

Comparative Performance of LOX/Kerosene Engines

	Aerojet-TRUD NK-31	P & W D-58N
Sea-level thrust (lb)	NA	NA
Vacuum thrust (lb)	90373	18700
Isp (sec) in vacuum	355	353
Chamber pressure	1334	1125
Mixture ratio, engine	NA	NA
Mixture ratio	2.6	2.6
Area ratio	114	184
T/W (calculated)	59	27
Engine cycle	staged combustion	staged combustion
Propellants	LO2/kerosene	LO2/kerosene
State of development	in production	in production

SOURCE: Ken Hampsten, "An Air-Launched, Highly Responsive Military TAV, Based on Existing Aerospace Systems," Briefing at RAND, January 22, 1996.

Linear Aerospike Engines

The linear aerospike engine has been the subject of research for many years. This type of engine has many potential advantages and is now planned as the main propulsion system for the X-33 technology demonstrator. The X-33 aerospike engine configuration is composed of small, side-by-side combustion chambers (thrusters) that exhaust onto a common exterior surface (nozzle), in contrast to the single combustion chamber exhausting into a bell-shaped nozzle found in traditional rocket engines. The general design and exhaust flow field for an aerospike engine are shown in Figure 4.3.

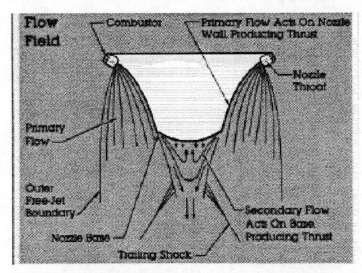

SOURCE: Aerojet Corp., *Aerospike*, Brochure, Pub.no., 671-M-87.6, 7 November 1987

Figure 4.3—Aerospike Engine Nozzle Operation

Exhaust gas expands against the nozzle on one side and into the atmosphere on the other side. The expansion characteristics of the propellant flowfield are influenced by the ambient pressure, since the outer surface of the flow is a free jet boundary.

At high altitudes, the free jet boundary expands to the Prandtl-Meyer turning angle. At low altitudes, the high ambient pressure compresses the primary flowfield, increasing the static pressure on the nozzle wall, as shown in Figure 4.4. Thus, the nozzle wall may be designed with a high area ratio for optimum high-altitude (or vacuum condition) performance without experiencing significantly decreased performance at low altitudes (high pressure condition). In contrast, a bell-shaped nozzle designed for high Isp in a vacuum would have an overexpanded jet, and relatively poor performance, at low attitudes.

There are some potential drawbacks to the aerospike engine. The aerospike engine has been found to exhibit a thrust loss through the transonic regime. Nevertheless, overall aerospike engine performance is superior to that of a conventional bell nozzle equipped rocket engine. There are also some concerns regarding vehicle stability during the ascent flight regime at max Q or when the vehicle experiences maximum dynamic pressure. The thrust vectoring control scheme (see discussion below) will have to be designed to accommodate stressing reaction control dynamics during this period.

Performance characteristics of the J2S aerospike engine planned for the X-33 are included in Table 4.3. It will have a relatively low T/W of 35 because relatively old gas generators (composed of older materials) will be used from the Saturn J-2 upper stage (from the Apollo program). Note, however, the low chamber

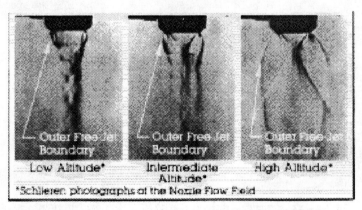

SOURCE: Aerojet Corp., *Aerospike*, Brochure, Pub.no., 671-M-87.6, 7 November 1987

Figure 4.4—Aerospike Engine Flow Fields

Table 4.3

J2S Aerospike Engine Performance

Sea-level thrust (lb)	202,480
Vacuum thrust (lb)	270,045
Chamber pressure (psia)	883
T/W	35
Mixture ratio, engine	5.5
Mixture ratio	5.84
Area ratio	87
Propellants	LOX/LH2
Total propellant flow rate (lbm/sec)	608.71

SOURCE: J. D. Elvin, "X-33 Ascent and Reentry Trajectory Analysis," Lockheed Martin Skunk Works, interdepartmental communication.

pressure of 883 psia planned for the X-33. The full-scale aerospike engine planned for the full-scale Lockheed-Martin RLV, the RS-2200, would have a higher chamber pressure of 2,250 psia, still significantly less than that of the SSME.

Aerospike engines are designed to be modular. Several modular units can be arranged in a specific configuration to form a propulsion system that can accommodate the shape of the vehicle and therefore minimize drag and optimize other aerodynamic performance characteristics. Several modular aerospike engine configurations are shown in Figure 4.5. On the X-33, the aerospike engine has two to three modules located side-by-side, each module having thrusters on opposite sides of a central truncated spike. In addition to fitting the rear profile

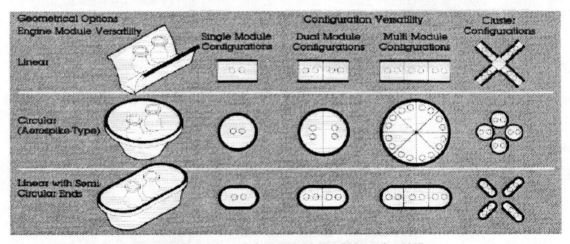

SOURCE: Aerojet Corp., *Aerospike*, Brochure, Pub.no., 671-M-87.6, 7 November 1987

Figure 4.5—Aerospike Engine Configurations

of the vehicle, this configuration allows thrust vectoring by throttling individual thrusters in the different modules, eliminating the need for mechanical gimbals and all the components associated with traditional thrust vectoring.

Although the concept of the linear aerospike engine dates back to the 1960s, the engine has never been flown and no flight-weight engines have been tested. Lockheed Martin plans to begin flight testing of the engine in October 1997, using an SR-71 as a supersonic testbed. The results of this test series will be correlated with ongoing wind tunnel testing at various facilities to verify analytical predictions.

Hydrogen Peroxide (H_2O_2) Propulsion

As mentioned in Section 2, hydrogen peroxide (H_2O_2) has been proposed as either an oxidizer or a monopropellant for TAVs. It has the advantage of being noncryogenic and having a relatively high density. It is not, however, without drawbacks. H_2O_2 presents a storage problem because it may not be stable in some situations, especially if contaminated. It can slowly decompose and evolve oxygen gas. At high temperature (e.g., if heat transfer in flight occurs), contamination can lead to an explosion hazard. It has been argued that contamination could occur during in-flight transfer.[1]

The high concentration of H_2O_2 required for propellants makes the substance very different from peroxide found in the household. The concentrated form can burn skin and poses a fire hazard to many organic materials. The vapors could pose a problem if there were faulty design, maintenance, or operations.

Nevertheless, its high density may make it an attractive propellant choice for air-launched TAV concepts, if it can be used safely. Unfortunately, no H_2O_2 rocket engines are available today. A new engine development program would be required.

A more attractive high-density propellant combination for air-launched TAV concepts may be LOX and methane, because propellant handling and storage may be easier on board the aircraft. Unfortunately, no existing LOX/methane rocket engines are available today.

---

[1]An argument has also been made that LOX can be contaminated and lead to potential hazards. It is not clear how this risk compares to the complications of in-flight refueling of peroxide.

Materials and Structures

The development of advanced composite materials may provide the needed structural or thermal performance that were limiting factors in earlier TAV and SSTO concepts. These materials include carbon-carbon, boron carbon, and graphite epoxy composites; and titanium-based metal matrix, ceramic matrix, and copper matrix composites. These are useful for high-speed vehicle structures subject to air friction without the weight penalty encountered by materials previously used in aircraft and TAV programs, such as aluminum alloys and the nickel alloy Inconel. At the RAND workshop, Vincent Weldon of Boeing reported that Inconel in a Ti-Inconel structure drove costs up significantly, and suggested that titanium be given serious consideration as a honeycomb structural material in metallic TAV or X-33 designs.

The space shuttle load-bearing structure is composed of aluminum alloy. Metals may still be used in current concepts for some structures, such as lightweight aluminum-lithium for oxidizer tanks internal to the structure that would not have to carry high heat loads. Composite material might also be used for the oxygen tank, although in the past there have been concerns over leakage. Where possible, tank structures are generally designed as integral parts of the structure.

Key properties of some of these materials are shown in Table 4.4. Advanced composites like boron epoxy and graphite epoxy exhibit high strength and stiffness. These materials are not in general unidirectional; they tend to be stiff in only one direction and less stiff in others. These limitations can be overcome by using multilayer laminate construction techniques.

Perhaps their most important property for TAV applications are their high strength-to-weight ratios and stiffness-to-weight ratios compared to standard

Table 4.4

Material Properties

Material	Specific Gravity	Tensile Strength (GPa)	Tensile Modulus (GPa)	Specific Tensile Strength (GPa)	Specific Tensile Modulus (GPa)
Boron/epoxy	2.0	1.49	224	0.73	110
Graphite/epoxy	1.6-1.5	0.93-1.62	213–148	0.58–1.01	133–192
Aramid/epoxy	1.45	1.1.38	58	0.95	40
Glass/epoxy	1.9	1.31	41	0.69	22
Steel	7.8	0.99	207	0.13	27
Aluminum alloy	2.8	0.46	72	0.17	26
Titanium	4.5	0.93	110	0.21	24

SOURCE: B. C. Hoskins and A. A. Baker, *Composite Materials for Aircraft Structures*, AIAA Education Series, New York, NY, 1986.

metals. Boron epoxy has four times the stiffness of an equivalent weight of steel, and graphite epoxy is five times stronger than the same weight of L65 aluminum alloy.

An autoclave is an oven that is used to cure composite parts. Because autoclave systems can be expensive, the design of TAVs based on composite parts must consider the cost of fabrication systems needed during vehicle construction.

Thermal Protection Systems (TPS)

Rockwell's space shuttle orbiter is the only current example of a TAV. The shuttle's aerodynamic load-bearing structures are composed of aluminum and titanium alloys and can not withstand the severe thermal loads experienced on reentry into the earth's atmosphere. Thus, an external thermal protection system is needed to isolate these aerodynamic load-bearing structures, as well as the internal structures of the vehicle, from the thermal loads.

Early TPS were originally expandable materials that heated and ablated under friction to protect the underlying structure. The shuttle uses relatively delicate silica glass tiles over much of the vehicle surface. However, its TPS must be carefully inspected and repaired after each flight—a major factor limiting shuttle responsiveness. Consequently, this type of TPS would not be well suited for a military TAV for which highly responsive operations would be a requirement.

Thermal loading on a TAV is dependent on the vehicle flight trajectory, as well as the aerodynamic design of the vehicle itself. The temperature profiles for various TAVs have been obtained from several sources. A summary is given in Table 4.5.

The first two columns contain predictions obtained from a Hypersonic Technology report by the National Research Council (NRC) on the National Aerospace Plane (NASP) and from a briefing by Boeing on its proposed Reusable Aerodynamic Space Vehicle (RASV); the last column contains data

Table 4.5

Selected TAV Temperature Distributions

	NASP[1]	RASV[2]	Space Shuttle[3]
Nose, leading edges	1650-2200 C	980-1520 C	1260-1540 C
Lower fuselage/wings	N/A	650-980 C	980-1260 C
Upper fuselage/wings	N/A	480-650 C	320-980 C

SOURCES: 1: RASV briefing, Boeing Aerospace Company, no date; 2: National Research Council, *Hypersonic Technology for Military Application*, Washington, D. C., 1989; 3: D. Curry, *Space Shuttle Orbiter Thermal Protection Systems Design and Flight Experience*, First ESA/ESTEC Workshop on Thermal Protection Systems, Netherlands, May 5–7, 1993.

68

taken from a Rockwell report on space shuttle missions. Similar temperature profiles were predicted for the pending X-33 technology demonstrator, as shown on the Lockheed Martin website.

The predicted thermal load on the NASP is much more severe than that predicted for the RASV or measured on the space shuttle. The NASP, however, is an air-breathing vehicle with design parameters distinct from the other vehicles. Temperature profiles of the latter two vehicles are similar and could be similar to those of a military TAV with similar hypersonic characteristics.

A detailed profile of peak temperatures on the space shuttle is shown in Figure 4.6. The profile indicates where thermal loads are highest and shows that these areas are relatively small compared to the entire surface area of the vehicle. However, because of the TPS used on the shuttle, these thermal loads are not distributed over the surface of the vehicle (the ceramic tiles are good insulators), resulting in the sharp temperature gradients shown in the figure.

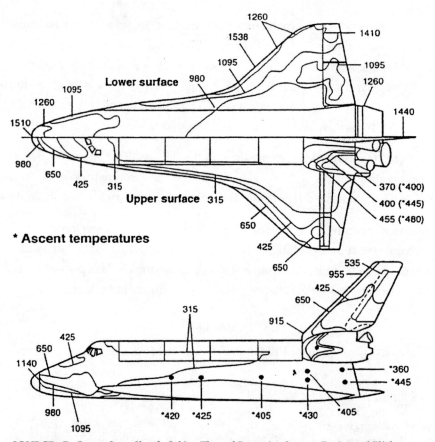

SOURCE: D. Curry, *Space Shuttle Orbiter Thermal Protection Systems Design and Flight Experience*, First ESA/ESTEC Workshop on Thermal Protection Systems, Netherlands, May 5–7, 1993.

Figure 4.6—Space Shuttle Peak Surface Temperature Profile

Current TPS Materials

A summary of the materials used on the space shuttle, the locations of their applications on the vehicle, and the maximum operating temperatures is given in Table 4.6.

Reinforced carbon-carbon (RCC) composites, having excellent strength and thermal properties, are used in peak thermal load regions. The RCC nose cap and wing leading edges also serve as aerodynamic load-bearing structures. They are attached to the shuttle frame via Inconel 718 and A-286 stainless steel fittings bolted onto flanges formed on the RCC components. The outer surface of the RCC panels are protected from oxidation with an silicon carbide coating. However, as Tables 4.5 and 4.6 reveal, on a typical mission the nose and leading edges of the space shuttle are subjected to temperatures exceeding the maximum operating temperature of RCC for 100-mission life. This suggests that the RCC panels have to be replaced frequently. On the space shuttle, these parts are indeed replaced after every mission.

The specific locations of space shuttle TPS components are shown in Figure 4.7.

Table 4.6

Space Shuttle TPS Materials

Material	Location on Shuttle	100-Mission Life	Single-Mission Life
Reinforced carbon-carbon (RCC)	Nose, leading edges	1482 C	1816 C
High temperature reusable surface insulation (HRSI)	Lower fuselage/wings	1260 C	1427 C (LI-900) 1482 C (LI-2200)
Fibrous refractory composite insulation (FRCI)	Selected HRSI regions	1260 C	1427 C
Low temperature reusable surface insulation (LRSI)	OMS[a] pod frontal area, area surrounding window panels	1093 C	1149 C
Felt reusable surface insulation (FRSI)	Upper fuselage/wings (low-temp regions)	399 C	482 C
Advanced flexible reusable surface insulation (AFRSI)	Upper fuselage/wings (high-temp regions)	816 C	982 C

SOURCE: D. Curry, *Space Shuttle Orbiter Thermal Protection Systems Design and Flight Experience*, First ESA/ESTEC Workshop on Thermal Protection Systems, Netherlands, May 5-7, 1993.

[a]OMS = Orbital Maneuvering System.

70

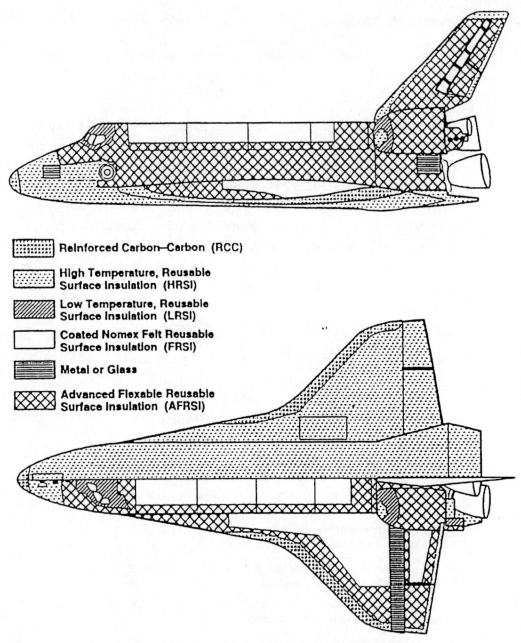

Reinforced Carbon–Carbon (RCC)

High Temperature, Reusable
Surface Insulation (HRSI)

Low Temperature, Reusable
Surface Insulation (LRSI)

Coated Nomex Felt Reusable
Surface Insulation (FRSI)

Metal or Glass

Advanced Flexable Reusable
Surface Insulation (AFRSI)

SOURCE: D. Curry, *Space Shuttle Orbiter Thermal Protection Systems Design and Flight Experience*, First ESA/ESTEC Workshop on Thermal Protection Systems, Netherlands, May 5-7, 1993.

Figure 4.7—Location of Space Shuttle TPS Components

HRSI and LRSI refer to ceramic tiles bonded to the aluminum skin of the space shuttle.

FRCI refers to a new ceramic composites that have improved strength and durability relative to HRSI tiles. FRCI tiles have been used on recent shuttle missions in place of HRSI tiles in areas susceptible to high impact damage. FRCI tiles have a higher energy density than HRSI tiles, however.

FRSI refers to flexible Nomex felt blankets used on the low-temperature regions of the shuttle body. These blankets are lighter than ordinary shuttle ceramic tiles, and require less time and labor to manufacture, inspect, and refurbish.

AFRSI refers to a new, silicon-based flexible blanket that has better thermal properties than FRSI. Because it is lighter and less complex than tile TPS, AFRSI has been used to replace a majority of the original LRSI tiles on the space shuttle.

Ceramic composites are excellent insulators and have a relatively low density, but have very low strength and ductility. Thus, it is necessary to isolate the ceramic insulation from aerodynamic stress loads. Independent motion of the tiles accommodates the flexing, expanding, and contracting motions of the metallic skin under aerodynamic and thermal loads. The main disadvantage of using ceramic tiles is system complexity (attachment to the aluminum substrate requires high-temperature adhesives, strain-isolation pads, filler bars, and precise gaps between the tiles). Ceramic tiles are also susceptible to impact damage because of the low strength of ceramic, and are water resistant to a temperature of only 570°C. The resulting TPS therefore requires rigorous inspection and refurbishment. Space shuttle inspection and refurbishment of the TPS require 17,000 man-hours per flight.

New TAV TPS Goals

To prepare the space shuttle for flight, months of inspection and refurbishment are needed between flights. Drastically increased responsiveness is desired for commercial launch operations, and is essential for military purposes. Plus, a reduction in mission costs would make transatmospheric missions more viable. These two goals imply that the following improvements must be made to future TAVs:

- Increased reliability: more durable and reliable equipment, requiring fewer inspections and refurbishments between missions.
- Increased simplicity: equipment requiring fewer parts and fewer man-hours for maintenance.

TPS need improvement for these new goals to be met. The disadvantages of the current shuttle TPS mentioned above must be overcome before significant advances can be made. The above TPS requirements can be met if materials can be manufactured that are stronger, lighter, have better thermal properties, and can be used as multi-function (i.e., aerodynamic and thermal load-bearing) structures. First introduced in the RASV program, advanced metallic alloy TPS may provide the reliability and simplicity needed for future TAVs.

Metallic TPS Materials

Metallic alloys offer several significant advantages over ceramic alloys. First, they are more resistant to impact damage. Also, they may be welded or fastened with rivets, potentially providing more flexibility in installation. Furthermore, because metals are conductive, they may redistribute heating loads over the body to reduce severe thermal gradients in peak heat-loading regions. Lastly, because they have greater strength and toughness than ceramics, metallic alloys can be used as aerodynamic load-supporting structures. This dual function of thermal and aerodynamic load support may eliminate the need for adhesives, strain-isolation pads, and gap fillers that are necessary with ceramic tiles. Lighter, more fragile internal insulation may also be used to enhance heat isolation.

The disadvantages of metallics are (1) higher density than ceramic tiles, (2) higher thermal conductivity, and (3) possibly greater difficulty in refurbishment if metallic panels are incorporated into load-supporting structures. The last disadvantage, however, is primarily a design problem associated with using the materials in a multi-functioning capacity.

Titanium aluminides are "advanced" metallic compounds developed for the NASP. The two classes of aluminides, alpha (TiAl) and beta (Ti_3Al), offer the advantages of a high maximum operating temperature ($800^\circ C$) compared to other titanium compounds ($500\text{-}700^\circ C$), oxidation and creep resistance, and, because of the high aluminum content, relatively low density. However, because the manufacturing processes have not yet been optimized, these materials have relatively poor room-temperature properties: they have low ductility, which results in low fracture toughness. In other words, they are more brittle than common titanium compounds at room temperature.

Commercially available advanced titanium alloys such as Ti-1100 and IMI-384 offer reasonable room-temperature ductility and toughness. They also possess high strength and creep resistance. However, a maximum operating temperature of $700^\circ C$ limits their range of usefulness as a TPS material. On the space shuttle, for instance, these titanium compounds may be used only on the upper fuselage. They also have higher densities than the titanium aluminides.

Nickel-based alloys such as Inconel 617 have significantly higher maximum operating temperatures (e.g., $1100^\circ C$ for Inconel 617), allowing use on lower fuselage surfaces and other high-temperature areas. They are also creep and oxidation resistant. They have significantly higher density than titanium alloys, however.

Performance-Enhancing Processes. Several manufacturing processes have been developed that allow "tuning" of material properties to enhance specifically desired characteristics. For example, heat treating enhances the strength of the material. Dispersion strengthening, in which particles of a different material are dispersed within the alloy matrix, enhances stiffness and thermal stability. These dispersion-strengthened materials are called metal matrix composites (MMCs). Finally, various coatings provide improvements in properties such as oxidation resistance, water resistance, and thermal emittance. These performance-enhancing processes allow the designer to improve thermal and structural characteristics of the material specifically where most needed.

X-33 TPS Materials. The Lockheed-Martin X-33 technology demonstrator uses a metallic TPS. According to representatives from Rohr (the subcontractor building the TPS), "the metal is more durable and lighter than normal shuttle insulation, but not as good an insulator." The lifting-body design of the X-33 allows reentry with a less severe heat-loading profile than the space shuttle, affording the use of the metallic TPS. In fact, temperatures on most of the bottom fuselage surfaces are predicted to be lower than 1300°F (700°C) and may be protected by titanium. Higher-temperature areas will have Inconel 617, and peak heating areas such as the nose and leading edges will be protected with RCC.

Several questions remain regarding the X-33 TPS. The first is the effect of the lifting-body design on the maneuverability and stability of the vehicle. Second, is the TPS fastened with rivets or bonded with adhesives? Are thermal sealants used in the gaps between the panels? Does the metallic TPS redistribute heating loads on the body surface? Also, are the RCC components designed to be replaced after every mission as they are with the space shuttle?

Observations on Thermal Protection Systems

Our review of TPS materials indicates that it should be feasible to design an alternative TPS from advanced metallic alloys, provided the reentry path and aerodynamic design of the vehicle body result in reentry temperatures that are less severe than those found on the space shuttle. Although peak temperature locations would probably still require RCC to withstand reentry thermal loads at selected areas like the nose, many TAV designs may be protectable by strategic combinations of metallic panels. Although metallic panels are of higher density than ceramic tiles, the metallic TPS may be lighter and simpler by eliminating the need for a complex adhesive system like that used on the space shuttle. The panels may also serve as aerodynamic load-bearing structures, eliminating the necessity for an underlying airframe.

Additional processing of the materials may enhance specifically desired properties to improve TPS performance.

Finally, by optimizing the vehicle's aerodynamic design it may be possible to reduce the thermal loads on the vehicle, thereby decreasing the degree of thermal protection required. These improvements and the other improvements mentioned above could result in a TPS that is more reliable and less expensive to maintain than that of the current space shuttle, thereby enabling development of a TAV that could responsively carry out military missions.

Vehicle Integration

To maximize orbital vehicle performance, it is necessary to efficiently integrate the engine, the airframe, and propellant tankage in a way that allows for good vehicle hypersonic characteristics as well as a minimum amount of added weight for subsystem integration interfaces (such as the TPS system and internal load-bearing structures or payloads).

If the system has two stages, they must be carefully and efficiently integrated as well. For either an aerial-refueled system or for air-launched system concepts, this implies careful attention to the mechanical interfaces and operational procedures during the staging maneuver. In either case, there is increased operational complexity and hazard introduced relative to an SSTO system.

Concepts in which the first-stage launch aircraft carry the TAV either under or on top of the aircraft must consider the drag effect that such structures have on the performance of the launcher platform, which must itself reach altitudes and speeds suitable for the system launch envelope. The separation dynamics of different TAV and aircraft combinations must be carefully considered, and for safety reasons subsonic air-drop configurations may be preferred.[2]

From the rocket equation for a SSTO design, typical delta-V required for orbit combined with an Isp of 460 seconds implies that 88 percent of the gross lift-off weight (GLOW) will be accounted for by propellant. SSTO and other TAV designs (including those using different fuels and lower specific impulse engines) must have a mass fraction sufficient to accommodate not only the

---

[2]The easiest approach is a subsonic air-drop, such as is used in Pegasus.

payload but also the structure, undercarriage and landing gear, propellant for delta-V maneuvering on-orbit, avionics, TPS, and other systems.

Mass fractions are more difficult to estimate for rocket vehicles than for conventional aircraft, especially because design experience for SSTO concepts is so limited. This introduces a significant uncertainty regarding weight of the mainframe and other results computed using aircraft-based models. Consequently, there is significant uncertainty regarding the TAV mass fractions predicted using these models.

Another vehicle integration and design issue is body shape, and the advantages or disadvantages of a design with regard to the encapsulation of propellant tanks and payload or crew areas. The role of wings versus lifting bodies must also be considered in terms of aerodynamics as well as weight goals. Lifting bodies have low lift-to-drag ratios (L/D) at subsonic speeds and moderate to high L/D at supersonic speeds, but do not have wings, which may introduce stability concerns on landing. On the other hand, wings are a parasitic weight when rocket propulsion is dominant over lift, as it is during the ascent phase. However, a design with small wings may imply high landing speeds, and this factor must be considered if the TAV design landing mode is horizontal.

Maneuvering in space is a fuel-consuming process, so the question is raised as to whether or not it is advisable to plan for a strategy of dipping into the atmosphere and using aerodynamics to achieve a cross-range capability. If this is done, then the effect on the lifetime of the vehicle must also be considered. Would a metallic design be more robust? What feedback effect would it have on demands for the propulsion system—would there be a significant savings over the life cycle of a vehicle if an atmospheric cross-range capability were built in?

Although LOX-LH2 offers high Isp, liquid hydrogen has a very low density, which in turn requires large tanks to contain it. This adds to the structural weight and volume of the vehicle. Such volumetric concerns are important system integration issues for air-launched TSTO TAV design concepts.

During the workshop sessions, it was indicated at the that a 1.5 percent weight ratio, significantly lower than for any other landing gear, was designed for the B-58 on display at the Wright Patterson Museum. The implication was that one could possibly trade landing gear weight for increased engine weight and perhaps gain increased propulsion performance.

TAV Design Scalability

Past experience with the design and development of aerodynamic vehicles demonstrates the value of starting with a subscale X (research) vehicle as part of an evolutionary development path toward a bigger, more capable Y (prototype) vehicle. An exploratory first step might be followed by an intermediate step to gather information on the effects of the environment, followed by a third step or expansion to the full-up operational vehicle. There were concerns expressed at the workshop that prototypes too frequently end up looking like the end item, and that the intermediate step is never really taken. The implication was that if R&D and user groups generally cannot agree on one design, vehicle design suffers and the potential scalability of the prototype vehicle design is reduced. Furthermore, if the need later develops for specialized vehicle applications, it may not be possible to accommodate those applications with a vehicle based on the original prototype concept. The capability to directly scale TAVs using subsystems that would be common and need only repackaging (i.e., no new designs) would be an ideal feature.

Special and unique scalability challenges apply to air-launched and aerial-refueled TSTO TAV design concepts. These challenges and their implications were discussed in the previous subsection and relate to the limits on the size of the orbital vehicle the first-stage carrier aircraft can carry to altitude, and practical limits as to how much propellant can be transferred to an orbital vehicle (and how quickly). Some and perhaps all of these scalability limitations can be overcome by developing new and larger transport or tanker aircraft, but this would raise the overall development for the design concepts significantly (anywhere from $2B to $5B for development alone of a new carrier aircraft) and may make them unattractive from a budget standpoint to SSTO RLV alternatives. Thus, it appears that air-launched or aerial-refueled TSTO TAV concepts would be attractive from a budget standpoint only if they are designed for small- to medium-sized payloads.

X-33 Vehicle Scalability Challenges

Two major scalability design challenges in going from an X-33 to an RLV that were mentioned by the working group were the need for the development of (a) high temperature composite thermoplastic types of materials and (b) extremely high thrust to weight ratio engines.

The discussion above on aerospike engines illustrates the scalability challenge faced by Lockheed Martin and Aerojet in the propulsion area. Similar propulsion scalability and performance challenges would face the other X-33 competitors in trying to scale up their vehicle designs from suborbital X-33 vehicles to full-scale SSTO RLVs capable of delivering sizable payloads to orbit.

5. Conclusions

TAVs could potentially launch payloads into space or toward distant targets at much lower cost than expendable launch vehicles. In addition, if they could be operated more like aircraft and less like expendable rockets, they offer the promise of carrying out space operations with much greater flexibility and responsiveness than is possible today.

Discussions at the workshop and subsequent investigation reveal that despite the efforts of past programs, significant technology challenges remain, especially in the areas of propulsion, thermal protection systems, and overall vehicle integration. Stringent mass fraction limits will have to be met for the vehicle to reach orbit with its intended payload. In this respect, overall vehicle design is very important. It is too early to know which sort of vehicle design has the best chances of meeting required mass fraction limits. More research is needed in propulsion, thermal protection systems, and overall vehicle design. The NASA X-33 program will provide important new data in all three areas, but the DoD needs to pursue research in all three areas as well.

A reusable launch vehicle could satisfy civil, commercial, and military space launch needs. However, our analysis reveals that civil and commercial launch needs differ in some important respects from emerging military needs. The highest priority for civil and commercial users is to reduce the cost of access to space. However, even though military users are also concerned about reducing costs, launch vehicle responsiveness and flexibility would be critical for some military missions. These differing needs would have an important bearing on vehicle design and imply that a military TAV may differ in important ways from an RLV designed exclusively for the commercial launch market.

And finally, reducing launch vehicle costs will at best address only half the problem of reducing the overall cost of access to space. Payload costs need to be reduced as well. Furthermore, there are subtle interactions between payload and launch costs. As launch costs increase, so do payload costs. To reduce the risk of on-orbit failure and the probability of relaunch, some payload subsystems are made triply redundant, increasing the cost and weight of the satellite. If launch costs can be reduced significantly, it may no longer be necessary to design in such high levels of redundancy. In addition, with TAVs it may be possible to recover payloads in orbit, and if payloads were designed modularly, they could be quickly repaired on-orbit. Such payloads could cost considerably less than existing satellites. TAVs may enable a new era of low-cost access to space.

Appendix

A. Agenda

Project AIR FORCE TAV Workshop

April 18

PAF Welcome RAND	8:30-8:45
Dynasoar Bill Walter	8:45-9:15
RASV/Have Region Dana Andrews, Boeing	9:15-10:00
Break	10:00-10:15
Role of TAVs/Spacecast 2020 Maj Chris Daehnick	10:15-10:45
U.S. Aerospace Forces 2020 and Technical Challenges Mike Snead, AFMC	10:45-12:00
Lunch	12:00-1:00
TAV Introduction and Technical Approaches/Issues Lt Col Jess Sponable, Phillips Laboratory	1:00-1:45
McDonnell Douglas X-33 Dr. William Gaubatz	1:45-2:45
Break	2:45-3:00
Black Horse Capt Mitch Clapp, Phillips Laboratory	3:00-3:45
Neptune Ken Hampsten, Phillips Laboratory	3:45-4:30
Reception Patio 6	5:00-6:00

April 19

X-34, X-33, and RLV Programs Overview William Claybaugh, NASA HQ	8:00-9:00
Lockheed Martin X-33 David Urie	9:00-10:00
Break for proprietary briefings	10:00-10:15

Rockwell X-33 Allan Lowry	10:15-11:15
X-34 Dr. Antonio Elias, Orbital Science Corporation	11:15-12:15
Lunch	12:15-1:15
Northrop Grumman Robert Haslett	1:15-2:15
TAV Presentation Lt Col Michael Baker	2:15-3:00
TAV Concept Vince Weldon	3:00-3:30
Lockheed Martin TSTO Ron Sullivan	3:30-4:15
Summary Discussion	4:15-5:15
Adjourn	

B. Participants

Project AIR FORCE TAV Workshop
April 18–19, 1995

NAME	ORGANIZATION
Charles Kelley	RAND
Glenn Buchan	RAND
Dan Gonzales	RAND
Mel Eisman	RAND
Calvin Shipbaugh	RAND
Bruno Augenstein	RAND
Tim Bonds	RAND
Bill Stanley	RAND
Dick Buenneke	RAND
Dr. Buzz Aldrin	Starcraft Enterprises
Mr. Dana Andrews	Boeing Defense & Space Group
Maj Brian Bergdahl	HQ USAF/XORR
Mr. Fred Bruckman	Lockheed Advanced Development Company
Mr. Raymond Chase	ANSER
Capt Mitch Clapp	Phillips Laboratory Reusable Launch Vehicle Office
Mr. William Claybaugh	NASA Headquarters Code XX
Maj Chris Daehnick	School of Advanced Airpower Studies
Mr. Carl Ehrlich	Rockwell International Space Systems Division
Dr. Antonio Elias	Orbital Sciences Corp.

Mr. John Fuller	Rockwell International North American Aircraft
Mr. Mark Gamache	Northrop Grumman B-2 Division
Dr. William Gaubatz	McDonnell Douglas Aerospace
Mr. Ken Hampsten	Phillips Laboratory Reusable Launch Vehicle Office
Mr. Robert Haslett	Northrop Grumman Advanced Technology and Development Division
Mr. Thomas Healy	Rockwell International Space Systems Division
Mr. Livingston Holder	Boeing Defense & Space Group
Lt Col Randy Joslin	HQ AFSPC/XPX
Mr. Harry Karasopoulos	WL/FIMA
Mr. John Livingston	ASC/XRE
Lt Col John London	Ballistic Missile Defense
Mr. Allen Lowry	Rockwell International Space System Division
Capt Tim Middendorf	Phillips Laboratory Propulsion Directorate
Mr. Joseph Nagy	NASP National Program Office MDC Office
Mr. H. Nichols	Phillips Laboratory Propulsion Directorate
Col Robert Preston	SMC/XR
Mr. Jim Pruner	WL/XP
Mr. Mel Rimer	Northrop Grumman Advanced Technology & Development Division
Mr. Tom Showers	HQ AFMC/DRS

Mr. Frank Snead	HQ AFMC/STPW Technical Requirements Division, Weapon Systems Branch
Maj Jay Snyder	AFSAA- Pentagon
Maj Greg Sparks	ASC/XRS
Lt Col Jess Sponable	Advanced Spacelift Technology
Mr. Ron Sullivan	Lockheed Martin Advanced Development Operations Hypersonic Systems
Mr. David Urie	Lockheed Advanced Development Company X-33 Program Manager
Col Chris Waln	SMC/XR
Mr. William C. Walter	Frontier Systems
Mr. Bill Warren	SMC/XR
Mr. Vincent A. Weldon	Boeing Defense & Space Group

References

Anselmo, Joseph C., "NASA Nears X-33 Pick," *Aviation Week & Space Technology*, June 17, 1996, p. 29.

Augenstein, Bruno, and Elwyn Harris, *The National Aerospace Plane (NASP): Development Issues for the Follow-On Vehicle, Executive Summary*, RAND, R-3878/1-AF, 1993.

Billig, F. S., "Design and Development of Single-Stage-to-Orbit Vehicles," *Johns Hopkins APL Technical Digest*, Vol. II, Nos. 3 and 4, July–December 1990.

Bunting, J., and S. Sasso, *The Success of the X-33 Depends on Its Technology—An Overview*, CONF 960109, American Institute of Physics, 1996.

Chase R. L., *A Comparison of Horizontal and Vertical Launch Modes for Earth-to-Orbit NASP-Derived Vehicles*, AIAA/SAE/ASME 27th Joint Propulsion Conference, AIAA 91-2388, June 24-26, 1991.

Curry, D., *Space Shuttle Orbiter Thermal Protection Systems Design and Flight Experience*, First ESA/ESTEC Workshop on Thermal Protection Systems, Netherlands, May 5-7, 1993.

Dornheim, S., "Follow-on Plan Key to X-33 Win," *Aviation Week & Space Technology*, July 8, 1996, p. 20.

Dotts, R., D. Curry, and D. Tillian, *Orbiter Thermal Protection Systems*, AIAA Minisymposium, New Orleans, Louisiana, September 23-24, 1983.

Eisman, M., and D. Gonzales, *Life Cycle Cost Assessments for Military Transatmospheric Vehicles*, RAND, MR-893-AF, 1997.

Elvin, J. D., *Lockheed Martin Approach to a Reuseable Launch Vehicle (RLV)*, CONF 960109, American Institute of Physics, 1996.

Elvin, J. D., X-33 *Ascent and Reentry Trajectory Analysis*, Lockheed Martin Skunk Works, interdepartmental communication, updated.

Gregory, Bowles, and Ardena, *Two Stage to Orbit Airbreathing and Rocket System For Low Risk, Affordable Access to Space*, NASA, April 1994.

Hampsten, Ken, "An Air-Launched, Highly Responsive Military TAV, Based on Existing Aerospace Systems," briefing at RAND, January 22, 1996.

Healy, T. J., Jr., *A Perspective on SSTO Systems*, Rockwell International, Space Systems Division, Downey, California (undated).

Hoskins, B. C., and A. A. Baker, *Composite Materials for Aircraft Structures*, AIAA Education Series, New York, 1986.

Hudson, Neff, and Andrew Lawler, "McPeak Presses for ASAT Options," *Defense News*, April 19, 1993.

Humble, R., G. Henry, and W. Larson, *Space Propulsion Analysis and Design*, McGraw Hill Inc. New York, 1995.

Mehta, U., *Air-Breathing Aerospace Plane Development Essentials: Hypersonic Propulsion Flight Tests*, NASA TM-108857, November 1994.

Moorman, Lt Gen Thomas S., Jr., *DoD Space Launch Modernization Plan*, briefing to the National Security Industrial Association (NSIA), 8 June 1994.

"NASA Gives Orbital Second Shot at X-34," *Aviation Week & Space Technology*, June 17, 1996, p. 31.

"NASA Nears X-33 Pick," *Aviation Week & Space Technology*, June 17, 1996, p. 29.

National Aero-Space Plane Materials and Structures Augmentation Program, *Titanium Aluminides/Advanced Monolithic Materials/Resin-Matrix Composites*, NASP/CR-1149, vol. 1, March 1993.

National Research Council, *Hypersonic Technology for Military Application*, Washington, D.C., 1989.

National Research Council, *Reusable Launch Vehicle Technology Development and Test Program*, Washington, D.C., 1995.

Penn, Jay P., *SSTO vs. TSTO Design Considerations—An Assessment of the Overall Performance, Design Considerations, Technologies, Costs, and Sensitivities of SSTO and TSTO Designs Using Modern Technologies*, The Aerospace Corporation, Space Technology & Applications International Forum (STAIF-96), January 7-11, 1996, Albuquerque, NM.

RASV briefing, Boeing Aerospace Company, no date.

Rockwell X-33 RLV User Expo, briefing, Downey, California, June 1996.

Ropelewski, Robert, "Boeing seeks to extend jumbo monopoly," *Interavia*, April 1995.

Sponable, Lt Col Jess, *Ground-Launched SSTO TAV Versus Air-Launched TAV*, Phillips Laboratory, PL/VTX, 2 May 1995.

Technical Requirements Document for a Military Trans Atmospheric Vehicle (TAV), Advanced Spacelift Technology Program, Phillips Laboratory, Space & Missiles Directorate, February 1995.

Wilson, A. (ed.), *Jane's Space Directory*, 11th Edition, UK, 1995.

Yost, Mark, "The Underground Threat," *The Wall Street Journal*, July 23, 1996, p. 22.